PHYSICS EXPERIMENTS AND PROJECTS FOR STUDENTS

Vol. II

PHYSICS EXPERIMENTS AND PROJECTS FOR STUDENTS

Vol. II

Edited by

C. Isenberg
University of Kent at Canterbury

and

S. Chomet
King's College London

Taylor & Francis
Publishers since 1798

PHYSICS EXPERIMENTS AND PROJECTS FOR STUDENTS

2 3 4 5 6 7 8 9 0 BCBC 9 8 7 6 5 4 3

ISBN 0-89116-270-4

Contents

Preface

Throughout the country there is a considerable number of undergraduate physics laboratories. The experiments in these laboratories have been developed by the local staff, often making use of the specialized interests of the department. There has been little interaction between academics who are responsible for these laboratories. Consequently I suggested to the Education Group of The Institute of Physics that it should organize an exhibition containing some of the standard teaching experiments available in undergraduate laboratories at universities, polytechnics, teachers' training colleges and institutes of higher education. This was subsequently extended to include advanced school projects. Such an exhibition would enable institutions to exchange information and ideas concerning experiments and enable new experiments to be introduced into laboratories without the considerable effort that is required to develop a new laboratory experiment.

The Education Group welcomed this suggestion. The first such exhibition was held in March 1984 and was a considerable success, attracting academic staff from all over Britain together with teachers and pupils from London schools. This success has encouraged us to organize future exhibitions of this nature and to publish this series of monographs.

This is the second volume in this series and it presents the experiments and projects presented at the second exhibition.

The third exhibition will be held at The Institute of Physics in September 1987.

CYRIL ISENBERG
University of Kent at Canterbury

GENERAL PROPERTIES OF MATTER

Solitons in a wave tank

N. Morton and J.L. Bingham
Department of Pure and Applied Physics, University of Salford, Salford M5 4WT

Abstract: Solitons are solitary, hump-like waves which are unchanged in shape as they travel, despite the anticipated distorting effects of medium dispersion. The distortion is exactly cancelled by the non-linearity of the system. In this experiment solitons are generated as surface waves on water in a wave tank. The wave profiles are detected by a wave monitor which measures the depth of water, the output being fed to a microcomputer and VDU for display. The launching parameters and the depth of water in the tank are varied to alter the number of solitons formed, and their amplitude and velocity between two spaced probes are recorded. The measurements are compared with theoretical predictions.

Introduction

The characteristics of waves excited on a one-dimensional system, such as a tensioned string, are very familiar. The waves exist due to the interplay between (i) the tension in the string T responsible for the restoring forces, tending to straighten the string and (ii) the inertia, represented by the mass per unit length of string m.

If the displacement y is small compared with a wavelength λ the string obeys the wave equation

$$v^2 (\partial^2 y/\partial x^2) = \partial^2 y/\partial t^2 \qquad (1)$$

where x and t are position and time, respectively, and v is the characteristic propagation velocity of a sinusoidal wave, equal to $(T/m)^{\frac{1}{2}}$.

Different solutions of the basic equation (1) correspond to different types of wave. For example, the standing wave shown in Fig. 1(a) may be represented by the function

$$y = a \sin (2\pi x/\lambda) \sin (2\pi \upsilon t) \qquad (2)$$

In general, three additional features exist for wave supporting systems:

Dispersion. The wave velocity varies with the wavelength or frequency, as for light waves in glass. A localised group or packet of waves may be regarded as a superposition of infinite, sinusoidal waves with a range of wavelengths. In a dispersive system the group will tend to change its shape as it propagates, for the various component waves travel at different speeds.

Non linearity. The basic partial differential equation equivalent to (1) contains terms which depend upon the displacement y to powers other

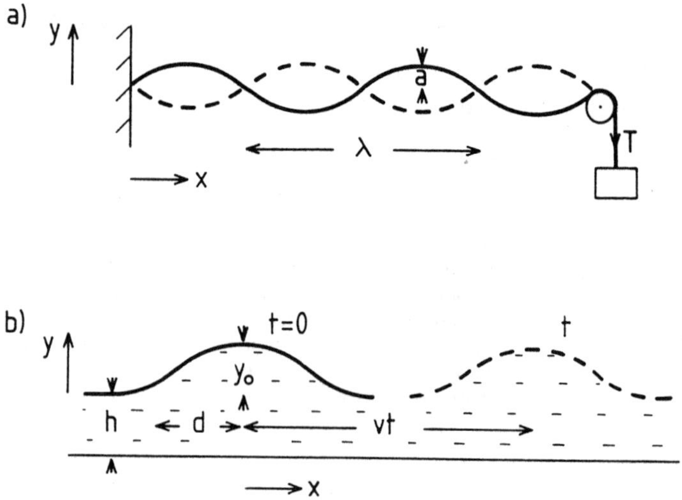

FIG. 1. (a) Standing wave on tensioned string; (b) soliton wave on surface of water ($d>h>y_o$).

than unity, resulting in a wave velocity which varies with the wave height or amplitude.

Friction. This progressively reduces the amplitude of the wave with increasing time.

In 1834, J. Scott Russell[1] noticed that a wave on the surface of water with only a single 'hump' (a solitary wave, now referred to as a soliton) could propagate along water channels with almost no change in shape over considerable distances.

The velocity of propagation varies with the height of the wave, the larger waves travelling faster. It is also possible to generate a short train of solitons. Waves on the surface of water exist due to the interplay between the restoring force of gravity, represented by the gravitational acceleration g, and the inertia of the water. Surface tension effects are often relatively small. Even for waves with very small amplitude, the wave velocity varies with the wavelength, i.e. the system is dispersive[2]. Korteweg and de Vries[3] investigated the mathematics of waves with wavelength substantially greater than the depth of water h, and with amplitude fairly small compared with h. They derived the following basic equation obeyed by the waves[3,4].

$$\partial y/\partial t + v_o \partial y/\partial x + (3v_o/2h) y \partial y/\partial x + \tfrac{1}{6} v_o h^2 \partial^3 y/\partial x^3 = 0 \qquad (3)$$

where $v_o = (gh)^{\frac{1}{2}}$ is the velocity of small-amplitude waves with wavelength far greater than the depth of water. The third term in (3) is non linear and the fourth term represents dispersion. It may be shown by substitution that a solution of (3) is a solitary wave with the hump-like shape shown in Fig. 1(b). The profile of the wave is given by[4,5]

$$y = y_o/\cosh^2 [(x - vt)/d] \tag{4}$$

where

$$\begin{aligned} v &= v_o(1 + y_o/2h) \\ d &= (4h^3/3y_o)^{\frac{1}{2}} \\ \cosh \theta &= \tfrac{1}{2}[\exp(\theta) + \exp(-\theta)] \end{aligned} \tag{5}$$

As the wave travels, its profile remains unchanged. The anticipated distorting effects of dispersion are exactly nullified by the non linearity.

Many systems are now known to exhibit soliton waves. For example, high-intensity light pulses travel in optical fibres as solitons, thus eliminating the inherent signal distortion due to the inevitable dispersion of light in the fibres.

Generation of solitons in a wave tank and principle of experiment

Water is held up behind the weir at the end of the wave tank, as shown schematically in Fig. 2(a). The height $2a$ and length $\frac{1}{2}L$ are recorded, and the weir is rapidly raised to launch the wave. It forms an initial, almost rectangular wave of height a and length L, as shown in Fig. 2(b). This rapidly forms a soliton or a short train of size ordered solitons. The total number N in the train is given by[4]

$$N = 1 + \text{integer part of } [(3a/2h)^{\frac{1}{2}}L/h\pi] \tag{6}$$

and their velocity is given by

$$v = (gh)^{\frac{1}{2}}(1 + y_o/2h) \tag{7}$$

so a graph of $v/(gh)^{\frac{1}{2}} - 1$ against y_o/h should have a slope of $\frac{1}{2}$. The validity of equations (6) and (7) can be investigated experimentally and the theoretical and experimentally determined wave profiles may be compared. The launching parameters a and L and the depth of water h in the tank are varied to alter the number of solitons formed, and their amplitude and velocity between two spaced probes are recorded. The wave profiles are detected by a wave monitor which measures the depth of water, i.e. the surface profile, by determining the electrical resistance betwen the pairs of wires in each probe. The wave monitor output is fed to a microcomputer, which runs the experiment, and to a VDU for display.

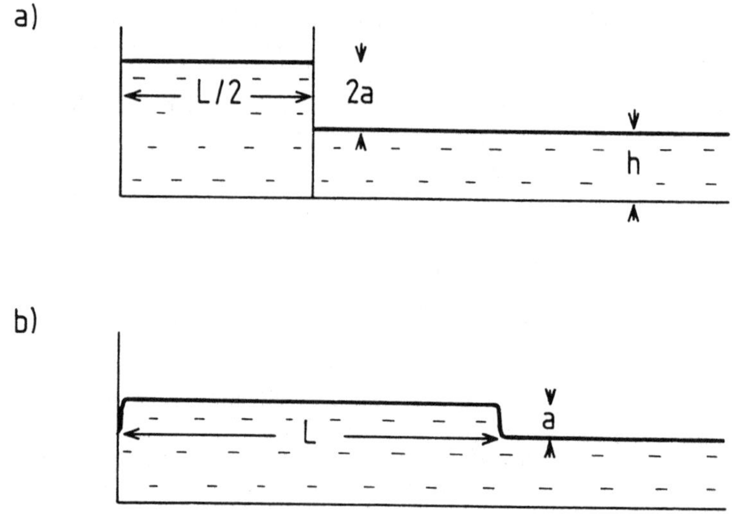

FIG. 2. (a) Soliton launching parameters: (b) initial wave.

Experimental details

The apparatus consists of a transparent plastic tank filled with tap water to a depth of a few cm. Attached scales allow the water depth to be measured. Values of $h = 6$ cm, $2a = 5$ cm and $\frac{1}{2}L = 20$ cm would be typical. A reasonably portable tank of length 2.5 m is adequate, but experiments in a longer tank are visually more appealing. A suitable weir may be made of two sheets of aluminium bolted together and nipping the inner limbs of a U-shaped, pliable plastic seal, furnished with a handle. In use it is pressed firmly against the sides and bottom of the tank. Water is piped behind the weir and it is raised to launch the solitons when a sufficient amount has accumulated. The program advises when initial conditions in the body of the tank are sufficiently stable. A 'beach' may be introduced at the other end of the tank to help damp out the waves, if desired.

The first probe is positioned sufficiently far along the tank (at least 1 m) so that the solitonms are well-developed by the time it is reached. The second probe is positioned a measured distance along the tank, and sufficiently distant from the far end to ensure that reflections do not interfere with the tail of the incident wave. The experimental arrangement is shown schematically in Fig. 3. The probes themselves comprise vertical, parallel stainless steel wires with a plastic spacer at the lower end, normally touching the tank bottom. The upper ends are gripped between

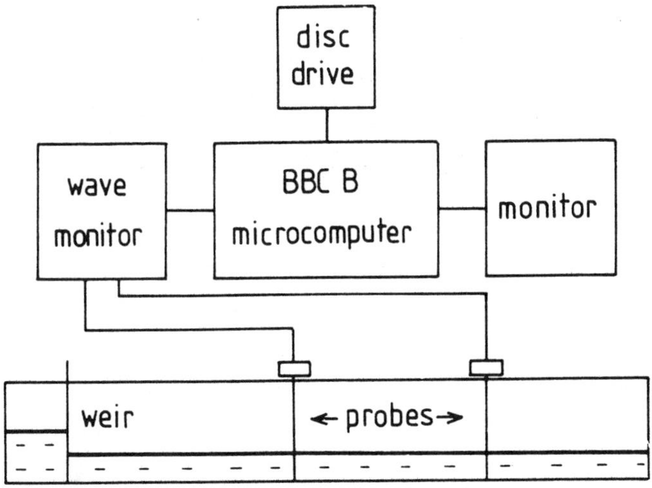

FIG. 3. Schematic diagram of apparatus.

grooved, plastic cross beams which support them in the tank. The probes are calibrated in an auxiliary flask or beaker following instructions given by the program. For this purpose the probes are raised in the water by slackening the grip of the cross beams. The twin channel wave monitor was supplied by Churchill Controls Ltd of Reading, England. A simple internal modification was made to it to ensure that only positive output voltages were supplied to the analogue-to-digital converter of the BBC B microcomputer.

Profiles recorded by the probes are displayed on the VDU monitor, and a moveable cursor in the display is adjusted to the wave peaks, allowing the wave heights and velocities to be recorded. Due to friction, the wave height reduces with time, more noticeably in a longer tank, so that the mean amplitude for the journey between the probes is recorded. Once the actual soliton velocity is known, the theoretical wave profile given by (4) and (5) may be compared with experiment. (It is most convenient to compute displacement y as a function of elapsed time by replacing x in (4) by vt_o, where t_o is the elapsed time corresponding to the wave peak. Alternatively, elapsed time may be converted to distance along the soliton.)

A listing of the program may be obtained from one of us (JLB).

REFERENCES

1 J. Scott Russell: *British Assoc. Reports* **6**, 417–496 (1838); **13**, 311–390 (1845).
2 J. Lighthill: *Waves in Fluids*, Cambridge University Press, p. 219 (1978).
3 D.J. Korteweg and G. de Vries: *Phil. Mag.* **39**, 422–443 (1895).
4 A. Bettini, T.A. Minelli and D. Pascoli: *American Journal of Physics* **51**, 977–984 (1983).
5 Lord Rayleigh: *Phil. Mag.* **1**, 257–279 (1876).

Forced oscillations and resonance

R.J. Cohen and D.A. Spinks
Schuster Laboratory, Department of Physics, The University, Manchester M13 9PL

Abstract: A simple experiment on the resonance response of a mechanical osillator to periodic forcing is described. The apparatus can be used to study many aspects of forced oscillations. A low-frequency phase meter used in the experiment is also described.

The phenomenon of resonance is central to most first-year undergraduate physics courses, yet interesting laboratory demonstrations of it are uncommon. MacLeod[1] has described a simple and elegant demonstration of resonance in a high-Q mechanical oscillator. This has been adapted and set up as a first–year experiment in our undergraduate teaching laboratory.

A photograph of our apparatus is shown in Fig. 1. The oscillator takes the form of a phosphor bronze bar which is rigidly clamped at one end. It is loosely coupled via a rubber band to a vibrator which is driven by an audio-frequency oscillator to provide periodic forcing. Velocity-dependent damping is provided by eddy currents generated in an aluminium plate at the free end of the bar, which moves between the poles of an electromagnet. A small coil which is attached to the underside of the bar moves between the poles of a permanent magnet, and generates a voltage which is proportional to the transverse velocity of the bar. The signal from the coil passes through a 20 dB amplifier to an oscilloscope. By measuring this voltage and the driving voltage, students can study the amplitude and

FIG. 1. Photograph of the apparatus.

phase response of the bar for diferent driving frequencies and different degrees of damping. The width of the resonance peak can also be compared with the decay time of free oscillations. The bar used has a resonant frequency of ~ 16 Hz and an undamped Q of ~ 500.

Phase measurements can be made using the oscilloscope in x–y mode to produce Lissajous figures, or using the twin beam display. However, many of our students prefer the simple phase meter illustrated in Fig. 2. The two sinusoidal driving voltages are used to generate switching signals which are fed to a zero-centred ammeter. If the input signals are proportional to sin ωt and sin $(\omega t + \gamma)$ the meter receives a positive voltage for $\pi - \gamma$ radians per cycle and a negative voltage for $\pi + \gamma$ radians. Therefore, provided that the meter has a suitable time constant, the meter reading is proportional to the phase difference γ between the two input voltages.

The apparatus is capable of yielding results which closely approximate the ideal behaviour of the simple damped oscillator treated in standard texts[2]. It also contains some non-standard features which students are invited to investigate. Several parameters of the bar can be altered, including the length of the cantilever, the point of application of the driving force and the coupling to the driving oscillator. The rubber band

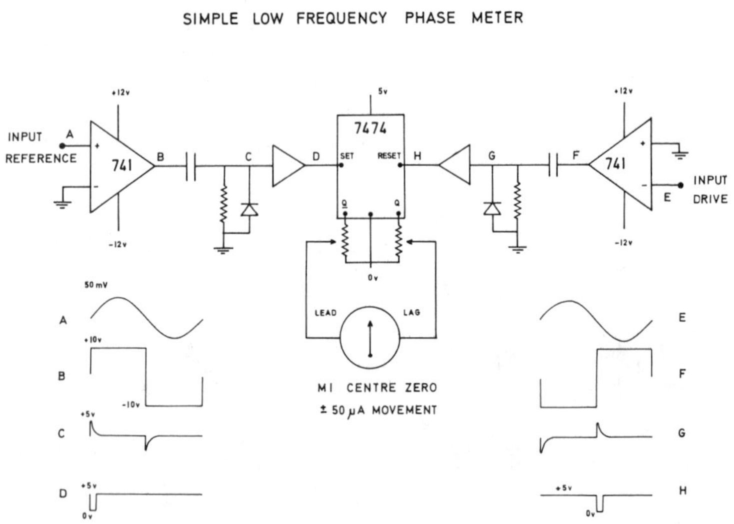

FIG. 2. Circuit diagram of the phase meter.

introduces a phase shift between the driving voltage (which is measured) and the driving force applied to the beam. It is also easy to produce non-linear behaviour by overdriving the beam. These and other features make the experiment popular with demonstrators, and usually ensure a lively laboratory interview with the students.

REFERENCES

1 A.M. MacLeod: *Physics Education* , **4**, 272 (1969).
2 A.P. French: *Vibrations and Waves*, Van Rostrand Reinhold (1982).

The not-so-simple pendulum

P.T. SQUIRE
School of Physics, University of Bath, Bath BA2 7AY

Abstract: A system for recording the amplitude of a rigid pendulum is described. The pendulum can be damped in a variety of ways, and the effect on the amplitude and period studied. Data can be recorded directly on a chart recorder for manual analysis, or by a computer for more detailed analysis. Several aspects of nonlinear, damped oscillations are conveniently demonstrated.

Apparatus

The pendulum is a fruitful source of teaching material from early school physics to final-year undergraduate courses. Many details of the motion are only apparent when precise records of the amplitude/time relationship are made. The demonstration apparatus described here is based on that of Nicklin and Rafert[1]. The pendulum is pivoted on a shaft, to which a servo potentiometer is attached. The potentiometer acts as a potential divider, such that the voltage across it is linearly related to the angular position. The voltage can be connected directly to a chart recorder; the trace then corresponds to the angle as a function of time.

Constructional details are shown in Fig. 1. The dimensions of the

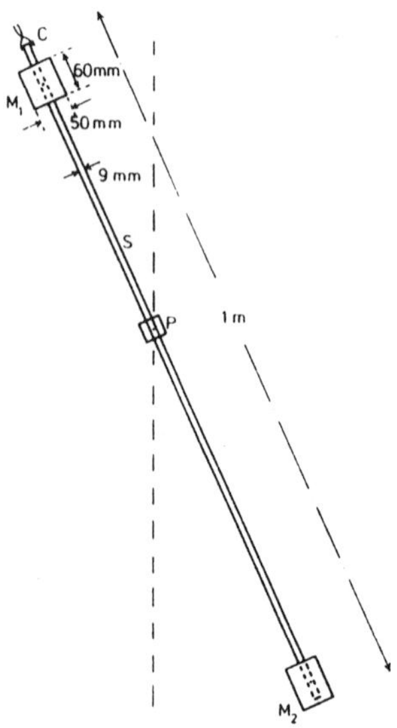

FIG. 1. Pendulum design.

pendulum are not critical; values given in the diagram are convenient. Two similar brass masses M_1 and M_2 are secured by screws to a brass shaft S. The masses are each about 1 kg, and the upper one can be slid along the shaft to vary the moment of inertia. A bulldog clip C is fixed to the top of the shaft to enable objects to be attached for studying air damping. The shaft is pivoted at P by means of a rectangular block. The block is attached to one end of a 6 mm steel shaft, which is supported by two self-aligning ball bearings. The other end of the shaft is attached to the servo potentiometer (RS Components No. 173-575). Provision is made for mounting circular metal discs at the front of the pivot for studying electromagnetic damping. Finally, a brake pad fixed to a hinged arm may be brought into contact with the horizontal shaft to provide dry friction damping.

The basic electrical circuit is shown in Fig. 2, where V_o is a stable d.c. voltage of about 5 V. The servo potentiometer forms two arms of a Wheatstone bridge, the other arms are variable resistors R_1 and R_2. The component values are not critical; for the potentiometer suggested, $R = 5\ \text{k}\Omega$, so that values for R_1 and R_2 of about 2.5 kΩ are appropriate. In a symmetrical arrangement the output voltage V is zero when the pendulum is vertical, but any asymmetry can be corrected by adjusting R_1 and R_2.

The amplitude is readily calibrated by setting the pendulum to known angles and noting the pen position on the chart. The scale can be quickly changed by using different ranges of the chart recorder or changing V_o, though overload of the potentiometer must be avoided.

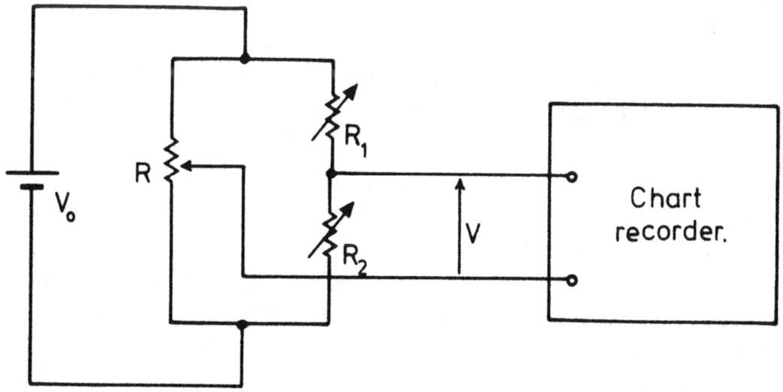

FIG. 2. Electrical circuit.

A typical trace for the nominally undamped pendulum is shown in Fig. 3. This shows that the decrement per cycle is less than 1%, mainly due to friction in the bearings and potentiometer. Note also the noise on the trace, manifest by the slight wobble on the envelope of the peaks, and seen most clearly by looking obliquely along the envelope. The cause of this is not known with certainty.

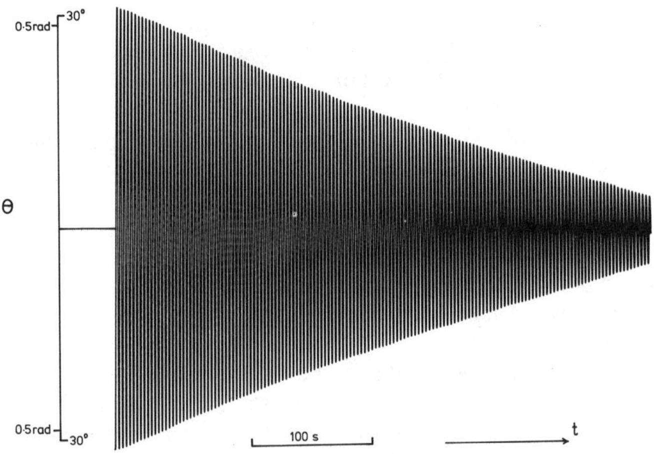

FIG. 3. Trace taken with residual damping only.

Mathematical model

The motion of a simple pendulum is commonly modelled by the simple harmonic oscillator equation

$$d^2\theta/dt^2 + (g/l)\theta = 0 \tag{1}$$

where θ is the angle between the pendulum and the vertical, l is the length and g is the acceleration due to gravity.

Although this equation is useful for revealing the basic periodic behaviour of a pendulum, it is too crude an approximation to a real pendulum to be satisfactory. A more realistic governing equation is

$$\begin{aligned} d^2\theta/dt^2 + a|d\theta/dt|d\theta/dt + b\,(d\theta/dt) \\ + c\,\mathrm{sgn}\,(d\theta/dt) + \omega_o{}^2(\theta - \tfrac{1}{6}\theta^3) = 0 \end{aligned} \tag{2}$$

which includes three damping terms. The first represents a damping force proportional to the square of the velocity, appropriate for air drag for all except very low Reynolds numbers. It makes the equation nonlinear, and, consequently, much more difficult to solve. The second term represents drag proportional to velocity, appropriate for low Reynolds number drag and for magnetic damping. The third term represents velocity-independent drag, appropriate to dry friction. ('sgn' denotes 'the sign of')

The restoring term $\omega_o^2(\theta - \frac{1}{6}\theta^3)$ contains the angular velocity ω_o for $\theta \rightarrow 0$. ω_o is given by $(dg/k^2)^{\frac{1}{2}}$ where d is the distance from the pivot to the centre of mass, and k is the radius of gyration. The restoring term also contains an approximation to sin θ that is useful for angles of swing up to 45°; this introduces further nonlinearity.

Equation (2) cannot be solved exactly, but approximate methods can be used to obtain formulae for the period and the amplitude. A_n denotes the amplitude of the nth swing, and $\Delta A_n = A_n - A_{n+1}$; that is, the decrease in amplitude from the nth to the $(n+1)$th swing. The formulae are

$$\Delta A_n = (8a/3)A_n^2 + \tfrac{1}{2}bT_nA_n + c(T_n/\pi)^2 \qquad (3)$$

and

$$T_n = 2\pi(k^2/dg)^{\frac{1}{2}}\,(1 + A_n^2/16 + \ldots) \qquad (4)$$

where A_n is the peak amplitude in radians.

It is assumed in deriving these formulae that the reduction in amplitude between successive swings is a small fraction of the actual amplitude. It is also assumed that the amplitude is no more than about 45°.

Damping studies

The three forms of damping corresponding to the three damping terms in equation (2) may be studied. The simplest approach is to make one term at a time dominant, so that each mechanism can be studied separately.

Dry friction (Coulomb damping)

A fibrous rubber pad is pressed against the rotating shaft with a controlled force. This produces a frictional force which, according to the usual model of dry sliding friction, is independent of the velocity. The result is a constant retarding torque that causes a constant angular deceleration. If the small change of period with amplitude [formula (4)] is neglected, this should cause the amplitude to decay by a constant amount each cycle. When air damping and magnetic damping are neglected, we have from (3)

$$\Delta A_n = c(T_n/\pi)^2$$
$$\approx c(T_o/\pi)^2$$

where T_o is the period for vanishingly small amplitudes. Thus a graph of A_n against n should be linear, with gradient equal to $c(T_o/\pi)^2$.

A typical amplitude/time graph is shown in Fig. 4. Direct inspection shows that amplitude does indeed decay very nearly linearly. Figure 5 shows the peak amplitude A_n plotted against n. Linear regression of the data for $0 \leqslant n \leqslant 28$ gives the following result for the Coloumb damping term in (3)

$$c = 0.0148 \pm 0.0004 \text{ rad s}^{-1}$$

The small curvature for $n > 28$ could be due either to other residual damping mechanisms or to a breakdown in the assumption of constant retarding force for very low velocities.

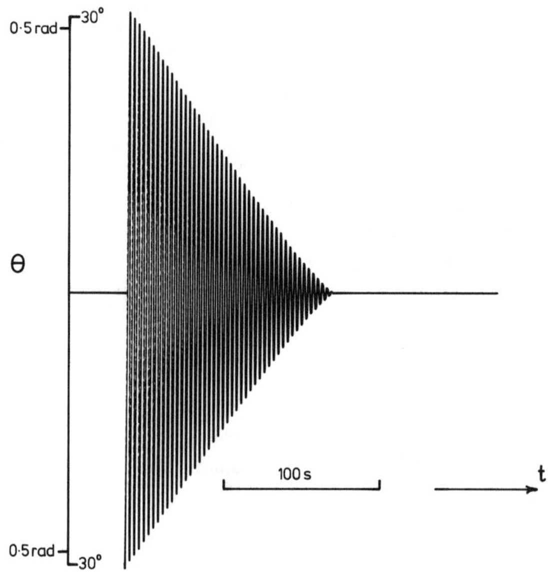

FIG. 4. Trace taken with dry friction (Coulomb) damping dominant.

Magnetic damping

An aluminium disc (2 mm thick, 30 cm diameter) is attached at the front of the pendulum by a screw coaxial with the horizontal shaft. A permanent magnet (Eclipse Major) is positioned so that as the pendulum oscillates the disc rotates through the gap of the magnet (see Fig. 6).

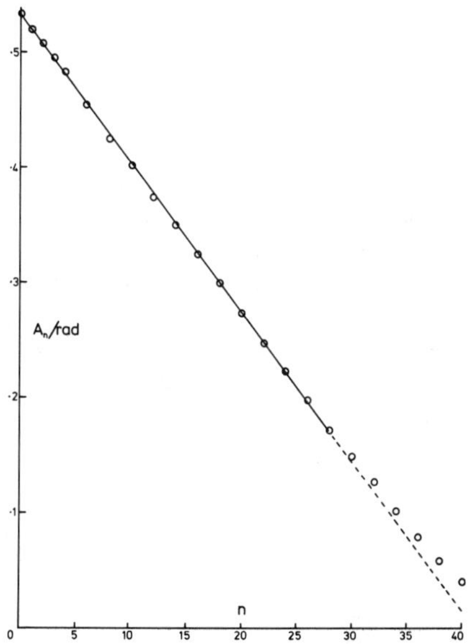

FIG. 5. Peak amplitude A_n of the nth cycle in Fig. 4.

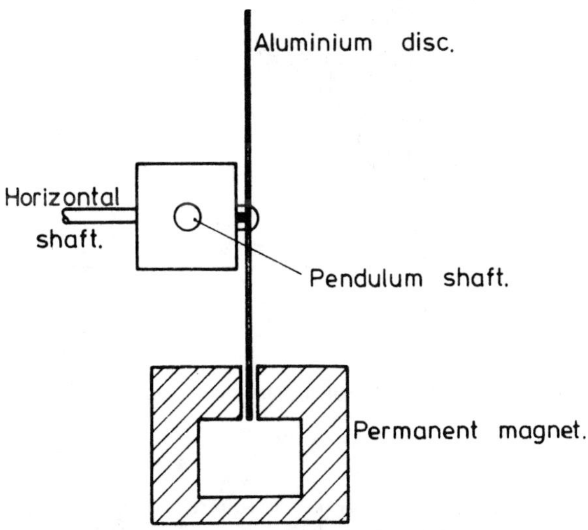

FIG. 6. Apparatus for studying electromagnetic damping.

Eddy curents induced in the aluminium disc are proportional to the angular velocity, and so the retarding torque in equation (2) is predominantly the term $b\mathrm{d}\theta/\mathrm{d}t$. Formula (3) then shows that

$$\Delta A_n = \tfrac{1}{2} b T_n A_n$$

Again neglecting the small variation of T_n with A_n we see that A_n decays exponentially with the number of oscillations.

Figure 7 shows a typical decay curve for magnetic damping, and Fig. 8 is a graph of peak amplitude against n on log/lin scales. It will be seen that the decay is close to exponential between 0.5 rad and 0.2 rad. Below 0.2 rad, residual friction causes the amplitude to fall off slightly faster. Linear regression of the data for $0 \leqslant n \leqslant 12$ gives

$$b = 0.0528 \pm 0.0004 \ \mathrm{s}^{-1}$$

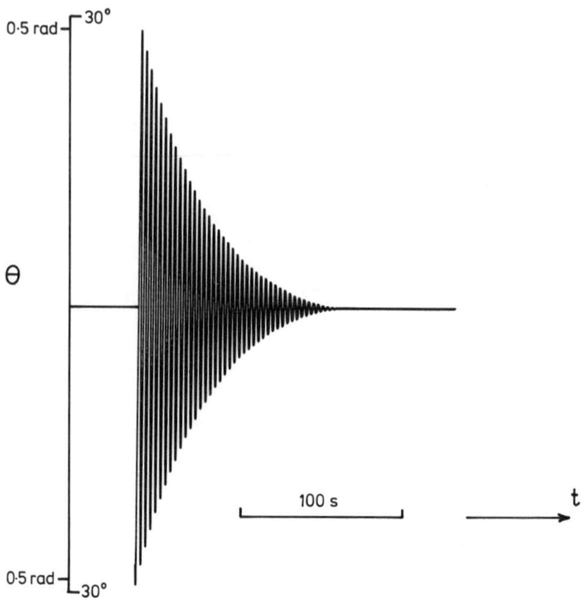

FIG. 7. Trace taken with electromagnetic damping dominant.

Air damping

A polystyrene tile approximately 0.25 m square is clipped on to the top of the pendulum. The mass is so small that the period is hardly changed, but if necessary the upper mass M_1 may be moved downwards to compensate.

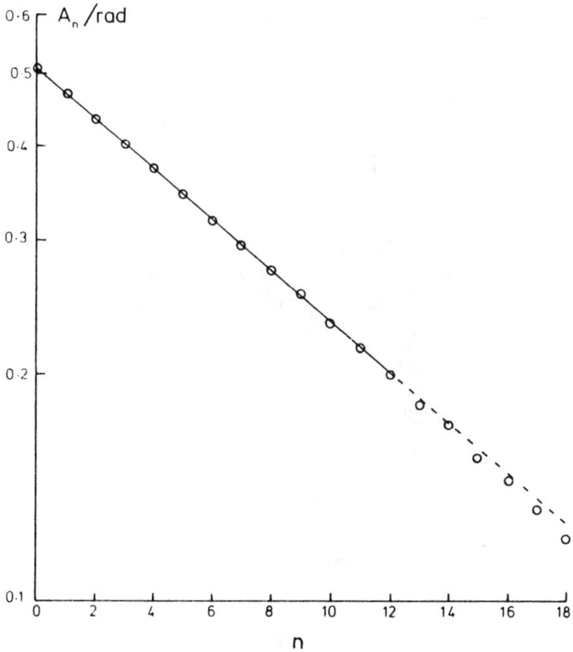

FIG. 8. Log/lin plot of peak amplitude A_n against n for Fig. 7.

The analysis of this case is more complicated than the previous two because the drag force does not depend in a simple way on velocity. For Reynolds numbers in the range $10^2 < \mathrm{Re} < 10^4$ it is a fair approximation to assume a constant drag coefficient. This leads to the quadratic damping term $a|\mathrm{d}\theta/\mathrm{d}t|(\mathrm{d}\theta/\mathrm{d}t)$ in (2). The rapid decay in this term means that the dry friction term cannot be completely ignored. Consequently, formula (3) takes the approximate form (for $A_n < 0.5$)

$$\Delta A_n = (8a/3)A_n^2 + c(T_o/\pi)^2$$

A graph of ΔA_n against A_n^2 should therefore be linear, with gradient $8a/3$ and intercept $c(T_o/\pi)^2$.

Figure 9 shows a typical decay curve, and Fig. 10 the graph of ΔA_n against A_n^2. The results are consistent with the assumptions made above, and linear regression of the data in Fig. 10 gives

$$a = 0.0422 \pm 0.028 \text{ rad}^{-1}$$

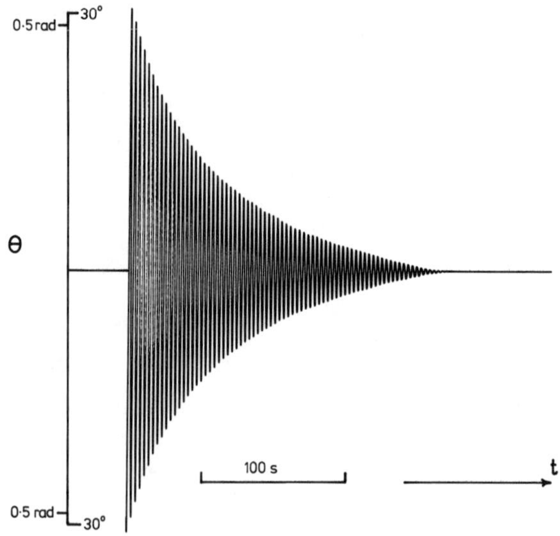

FIG. 9. Trace taken with air damping dominant.

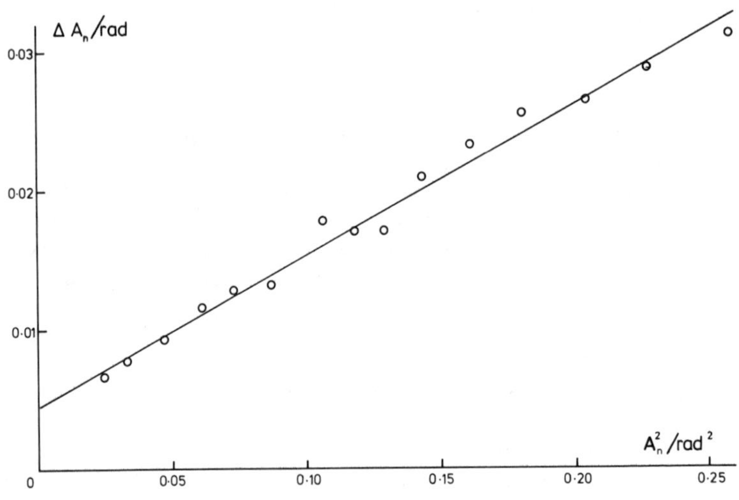

FIG. 10. Graph of ΔA_n against $A_n{}^2$ for Fig. 9.

Further developments

More detailed studies of damping, including two or more mechanisms simultaneously and the variation in period, can be made by logging the results directly into a computer equipped with a suitable analogue to digital convertor[1]. An instrument bus and suitably interfaced timer/counter and analogue to digital convertor are useful refinements.[2] However, the extreme simplicity of the basic apparatus and the rapidity with which it can produce graphic results make it attractive both in the laboratory and for lecture demonstrations.

REFERENCES

1 R.C. Nicklin and J.B. Rafert: The digital pendulum, *American Journal of Physics,* **52**, 632–639 (1984). See also S.C. Zilio, Measurement and analysis of large-angle pendulum motion, *American Journal of Physics,* **50**, 450–452 (1982).

2 P.T. Squire: Pendulum damping, *American Journal of Physics,* **54**, 984–991 (1986).

Simple measurement of apparent depth

E.G. HUBERT
Department of Physics, Mathematics & Computing, Manchester Polytechnic, Manchester, M1 5GD

Abstract: Many two-dimensional phenomena are perceived in depth. A simple optical system is used in this experiment to detect depth in the well known Muller-Lyer illusion.

The two-dimensional is every bit as fictitious as the four-dimensional, for nothing is flat, not even the most finely polished mirror — M.C. Escher (Graphic Artist) 1960.

Some two-dimensional figures have apparent depth. Indeed this is the basic illusion intended in most artistic pictures.

Perspective is often used to create such a depth effect. In Fig. 1, which shows the Ponzo illusion, the perspective places the upper rectangle further away than the lower. Why does the upper rectangle look longer? One theory states that those parts of illusion figures which would represent distant features are enlarged, whilst those representing near features are shrunk.

This may be in keeping with the effect of 'size constancy'. Brain scaling ensures that distant objects are scaled 'up' to maintain their perceived size. A person who is 6 ft tall close up still looks 6 ft at a greater distance away even though the retinal image formed by geometric optics is obviously smaller. Of course, in normal viewing we would not be able to

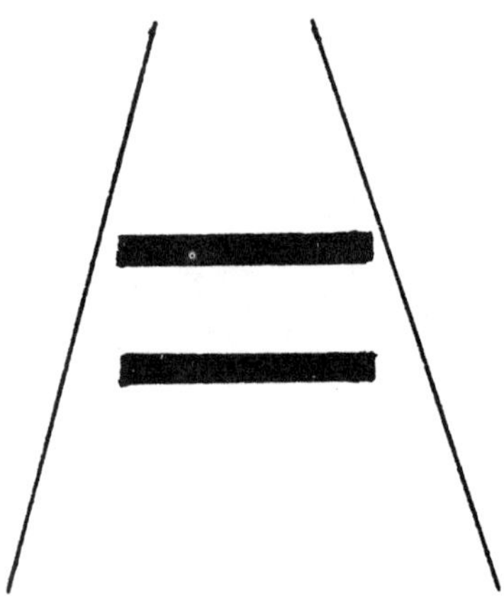

FIG. 1. Ponzo illusion.

separate retinal changes in image size from this brain scaling. However, if a good after-image is formed from, for example, gazing intently at the black dot in Fig. 2 for some time and then looking at a blank screen, dramatic variations in size can occur. If the attention is changed from a blank sheet at arm's length to a distant wall the instantly recognised after-image will appear much larger. The retinal image stays the same but the brain scaling now works to increase the perceived 'object' size as its distance away is increased (Emmert's Law). We can therefore conclude that in normal object viewing the scaling up occurs with the image size perceived by the eye plus brain system.

FIG. 2. Figure to produce a famous after-image.

All of this depends on there being apparent depth in the picture. Can this depth be measured? Gregory[1] devised the apparatus shown in Fig. 3 to do this. A moveable spot is viewed binocularly in order to permit depth judgements to be made, whilst the object is viewed monocularly. Light from the spot enters both eyes, having in one case travelled with little noticeable effect through a single sheet of polaroid. Crossed polaroids prevent one eye from seeing the object which is back illuminated to avoid the creation of any 'false' depth effects from texture.

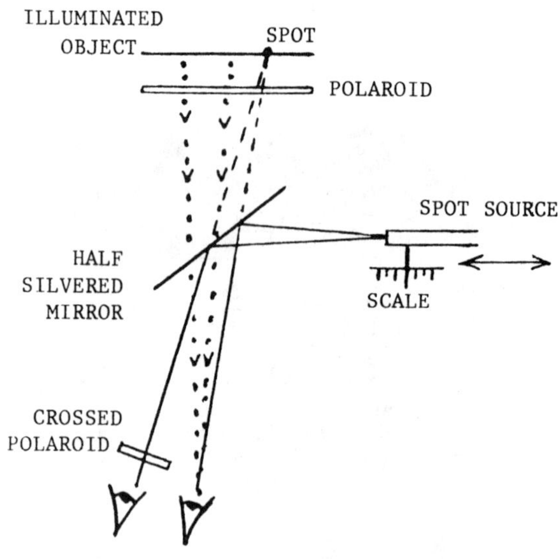

FIG. 3. Plan view of apparatus.

The spot (or spots, if necessary) can be moved to be alongside any feature of the picture. Any appearance of depth can therefore be recorded on the scale alongside the spot source. A most appropriate figure to test for apparent depth is the famous Muller-Lyer figure (Fig. 4). The left hand (fins in) picture looks smaller than that on the right (fins out). One theory is that the fins via their perspective features lead us to see the shape in three dimensions. In Fig. 4 the illusory figures are shown alongside normal day to day viewing events with similar features. If the theory discussed earlier holds then the near corner would be shrunk and the far corner enlarged.

In the experiment, a series of different angle fin shapes of the same shaft length are viewed in turn at a fixed distance from the illuminated

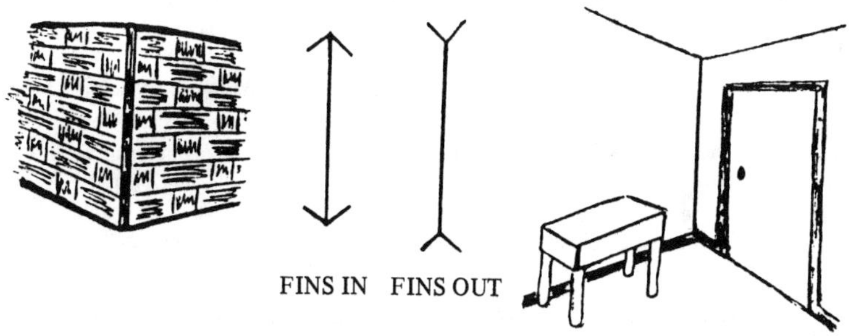

FIG. 4. Muller-Lyer illusion alongside everyday counterparts.

object screen. It is important that the object size is set so that the whole arrow figure is viewed. The different positions of the spot (as measured on the scale) should then indicate the apparent depth in the figures. Gregory[1] found a relationship between the apparent depth so measured and the apparent length (measured separately) of the Muller-Lyer figures.

The apparatus can also be used to demonstrate apparent depth in other two-dimensional figures, e.g. the Ponzo illusion, different coloured shapes on a background, artistic effects and so on.

Dimensions and other details of the apparatus can be obtained at the above address from the author or Mr. S. Appleby (Laboratory Superintendent, Manchester Polytechnic) who transformed a paper description into a compact working reality for use by students.

REFERENCE

1 R.L. Gregory, *Eye and Brain.* Weidenfeld & Nicholson, 1979.

Psychophysical measurement of area

E.G. HUBERT
Department of Physics, Mathematics & Computing, Manchester, M1 5GD

Abstract: Physics students are usually encouraged to measure variables as accurately as possible with the most suitable device available. This experiment forces students to use their own visual system to measure the areas of circles on a computer screen. The relevant psychophysical theory is then developed.

"A real measure of sensation would demand that we be able to call a sensation twice, thrice or so-and-so many times as intense as another — but who would say such a thing?" G.T. Fechner 1860

Instruments such as rulers, verniers and micrometers are used in physics to measure length, area and so on, more accurately than is possible with the unaided human eye. However, in making such judgements in everyday experience we often have no alternative other than the direct use of our visual system, i.e. eyes plus brain. How accurate is this mechanism? Can we predict how different people would make the same measurement?

If the sensation to a given stimulus cannot be measured directly then we would need to record the number of just noticeable differences (JNDs) above the minimum perceptible level (or absolute threshold) that are required to build up to the perceived sensation. Weber's law states that a JND grows larger in direct proportion to the size of the stimulus. Fechner further proposed that each time a JND is added to the stimulus, the sensation increases by a jump of constant size.

Thus we have the Weber-Fechner law

$$\Delta x/x = \Delta y$$

where x is the stimulus (actual length, area, sound etc. presented) and y is the sensation (personal reaction to the stimulus). This can also be expressed as $y = \ln x$.

The indirect measurements involved are based on differences. However, many of the judgements made in everyday life are based on ratios rather than on differences. Stevens[1] and others have attempted direct measurements of various sensations such as brightness, loudness, taste and so on. In these measurements, known as magnitude estimations, care was taken to avoid judgements based on difference because these led to 'Fechner like' partition (or category) scales. A power law has been proposed for such magnitude estimations, namely, $y = ax^b$ or $\ln y = b \ln x + \ln a$. The constant a depends on the units of measurment and is of limited interest. However the exponent b varies from around 0.5

for brightness to 3.5 for an electric shock through fingers. It is not proposed to test the latter!

Instead, this experiment tests these predictions from direct estimates of the areas of seven circles presented on a computer screen. Filled-in red circles are used to encourage perception as discs and the measurement of area rather than diameter.

A middle sized area is presented first and the observers are told that subsequent presentations will involve both larger and smaller circles. This encourages judgements based on ratios rather than on difference. Thereafter the circles are presented in a random order decided in advance by the viewer.

Computer presentation appears to have several advantages over the cards used by Stevens[1]. The time of presentation is pre-determined and kept constant and the program waits for the observer to enter his estimate before showing the next circle. Instructions are given in an easily assimilated form and include the statements that the observer should try to estimate the apparent size directly and not attempt to calculate the area from the perceived diameter. Powers of calculation are not being tested.

Finally, a built-in program accumulates the results and on completion gives a plot of the points obtained together with the closest power-law curve fit. The calculated exponent is then displayed together with the estimated error involved. The latter is an useful indicator of those 'non-believers' who have not made a serious attempt at the estimation! Without the curve fit facility the results could be recorded and a log–log graph plotted to determine the exponent from the slope of the 'best' line.

The anticipated value for the exponent is 0.7. In a demonstration of the experiment the mean result from 50 observers was 0.8. A typical result is shown in Fig. 1.

Although this is not essential, the circles can be chosen to include a mid-point stimulus, an anticipated mid-point magnitude estimation and a mid-point logarithmic law sensation, thereby readily demonstrating the principles and contrasting laws involved. A difference measurement can be induced by presenting the smallest circle first, followed by the largest and the others at random. Stevens[1] gives various predictions on the category scale then involved in relation to the magnitude scale. The effect of a different shape and/or colour could also be investigated.

A listing of the program as written for the BBC microcomputer can be obtained from the author at the above address.

Acknowledgements

Miss. R. Burrell and Miss. A. Vastano (Combined Studies students in Applied Physics and Computing Science) translated a simple program

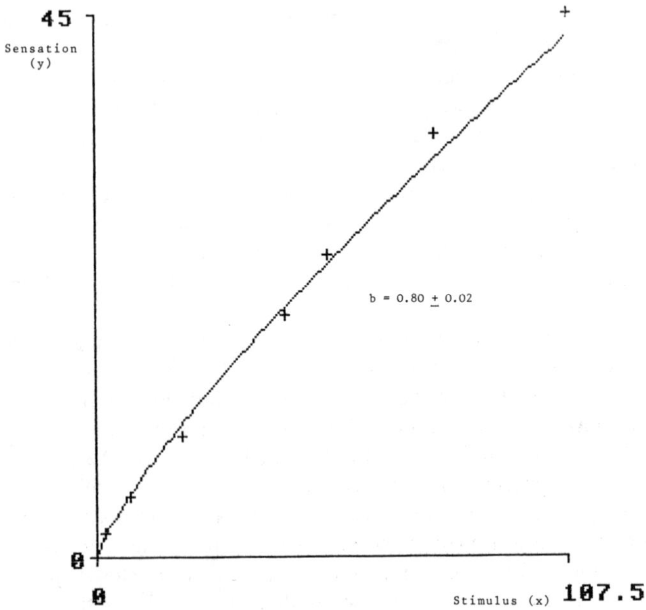

FIG. 1. Typical results for experiment.

into BBC BASIC and improved the presentation and Mr. J.E. Hoad (Senior Lecturer, Manchester Polytechnic) adapted his curve fit program to link into the package.

REFERENCE

1 S.S. Stevens: *Psychophysics*. John Wiley & Sons (1975).

LIGHT AND SOUND

An introduction to Fourier transform spectroscopy

E. WEBSTER AND P.A. BATES
Lancashire Polytechnic, Preston, Lancashire PR1 2TQ

Abstract: A simple computer controlled Fourier transform spectrometer is described. The experiment under the control of an Apple II microcomputer generates interferograms and calculates the associated spectrum using a fast Fourier transform program.

Fourier transform spectroscopy has for a number of years been a valuable technique in the analysis of infrared spectra but because of the need for dedicated computing facilities it has been confined to the research laboratory. The increased availability of small computers such as the Apple II and BBC and the design of fast Fourier transform programs in Basic[1,2] for these machines enables a simple experimental package to be produced for the undergraduate laboratory.

If a Michelson interferometer is arranged in the Twyman Green form so that the mirrors are parallel, Fig. 1, where M_2^1 is the image of M_2 then the path difference for radiation of wavelength λ is

$$\varphi = (2\pi/\lambda)2h$$

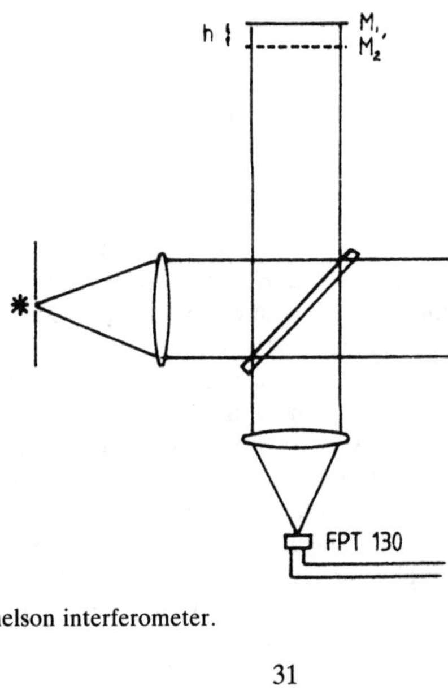

FIG. 1. Michelson interferometer.

The two beams recombine at the semisilvered miror and if they have the same amplitude A then the resultant intensity is given by the equation

$$I = A^2 + A^2 + 2A^2 \cos \psi$$

$$I = 2A^2 [1 + \cos (2\pi 2h/\lambda)]$$

It is often better to work in terms of wave number $\sigma = 1/\lambda$ and to replace $2h$ by the path difference l in which case

$$I(l) = 2A^2(1 + \cos 2\pi\sigma l)$$

When the incident light is not monochromatic but contains a range of wavenumbers σ then the resultant intensity is the sum due to each component

$$I(l) = \int_0^\infty B(\sigma)(1 + \cos 2\pi\sigma l)\, d\sigma$$

$$I(l) = \int_0^\infty B(\sigma) d\sigma + \int_0^\infty B(\sigma) \cos 2\pi\sigma l\, d\sigma \qquad (1)$$

When the two mirrors are equidistant

$$I(l) = I(0) = 2\int_0^\infty B(\sigma)\, d\sigma$$

$$\int_0^\infty B(\sigma)\, d\sigma = I(0)/2 \qquad (2)$$

Substitution of (2) in (1) gives

$$I(l) - I(0)/2 = \int_0^\infty B(\sigma) \cos (2\pi\sigma l)\, d\sigma$$

which is a cosine Fourier transform relationship from which one can write

$$B(\sigma) = \int_0^\infty (I(l) - I(0)/2) \cos 2\pi\sigma l\, dl \qquad (3)$$

This equation is the basic equation of Fourier transform spectroscopy since it enables the intensity spectrum $B(\sigma)$ of a source to be obtained for each value σ by performing the integration in equation (3).

The practical steps taken to obtain $B(\sigma)$ are as follows:

(1) $I(l)$ is obtained as a function of l using the Michelson interferometer and this is known as an interferogram (Fig. 2).

(2) The interferogram is related to the visibility curve $[I(l) - I(0)/2]/I(0)/2$ and so $B(\sigma)$ can be derived from the Fourier transform of the visibility curve derived from the interferogram.

(3) The magnitude of $B(\sigma)$ for each wavenumber, σ, say, is obtained by multiplying the whole of the curve $[I(l) - I(0)/2]$ by $\cos 2\pi\sigma l$ and finding the area under the curve. This is performed by the computer using an algorithm known as a fast Fourier transform.

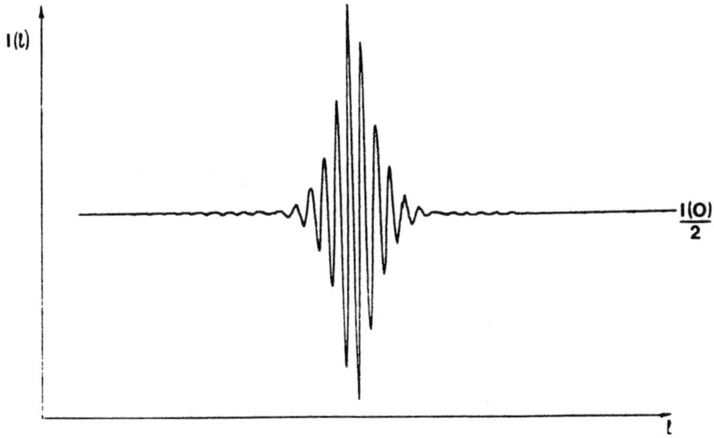

FIG. 2. White light interferogram.

The practical apparatus uses a simple Beck Michelson interferometer[3,4] with the addition of a stepping motor to drive the micrometer using a friction cord pulley. An Apple II microcomputer was used both to record the data and drive the motor using the A and B side of a single PIA placed in slot 3. The spectrum was detected using a Fairchild FPT 130 photo-transistor, the output from which was amplified using a single stage operational amplifier with variable gain to generate voltages in the range 0–5 V. This analogue signal was then fed to a National Semiconductor ADC 0804 LCN analogue to digital converter wired to run in the continuous conversion mode (Fig. 3). The B side of the PIA was used to drive a small three-phase stepping motor via a Darlington driver circuit, Figure 4 shows the complete layout of the system.

With the interferometer set to a uniform intensity background the output from the operational amplifier was adjusted to give a 2 V signal and the demonstration program supplied measured 1024 data points across the interferogram using signal averaging to reduce the noise.

Lines 100–260	Sets up the PIA and generates suitable codes to drive the stepping motor.
Lines 270–430	Reads the intensity data and stores the information as an array RE(X).
Lines 440–640	Plots the interferogram on the VDU produces a paper print of the spectrum and stores the data as a file on disc.

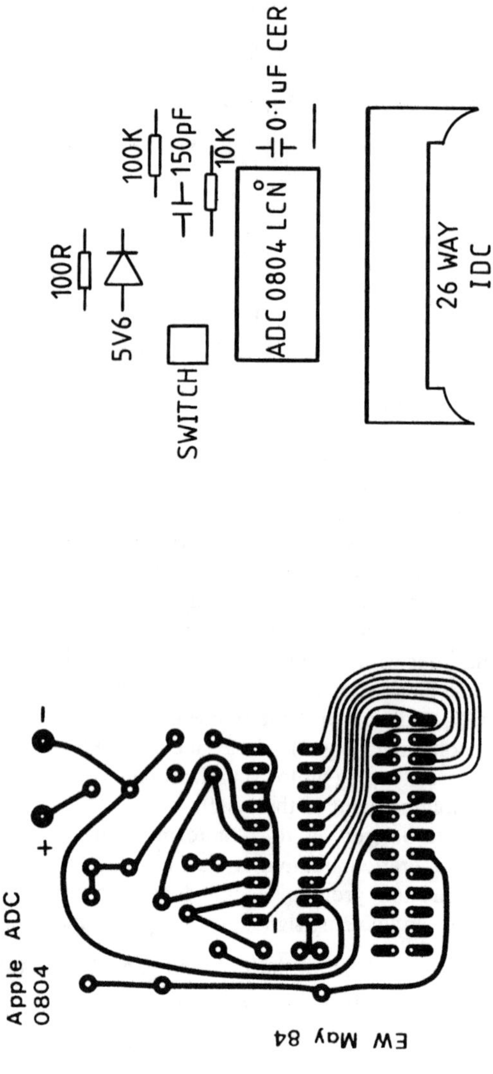

FIG. 3. Printed circuit board for ADC 0804 LCN.

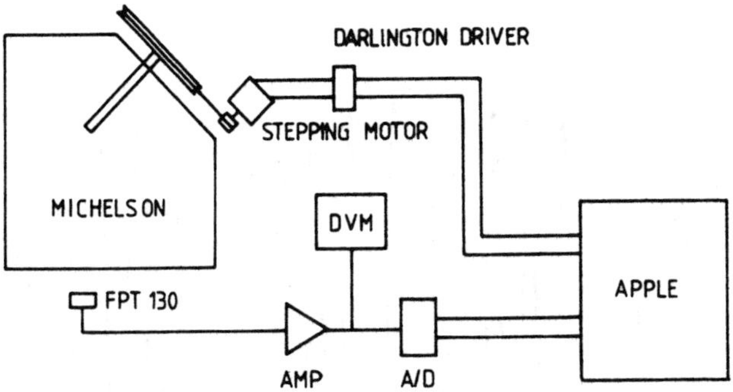

FIG. 4. General component layout.

Lines 650–930	Generates the spectrum for 1024 points, plots the spectrum on the VDU and produces a paper print.
Lines 950–1100	Saves the spectrum on disc and rewinds the interferometer to repeat the measurement.
Lines 1120–1450	Is a subroutine used in the FFT program.

Figures 3, 5, 6 and 7 show the interferograms and spectra for a white light and laser source plotted on a digital plotter from data stored on disc by the program (Tables 1, 2 and 3).

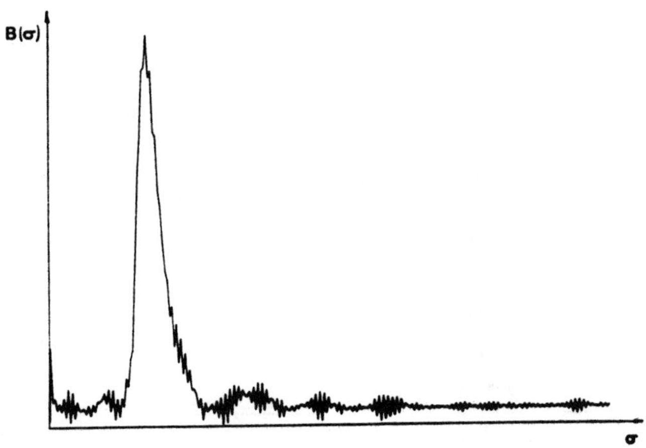

FIG. 5. White light spectrum obtained from the white light interferogram.

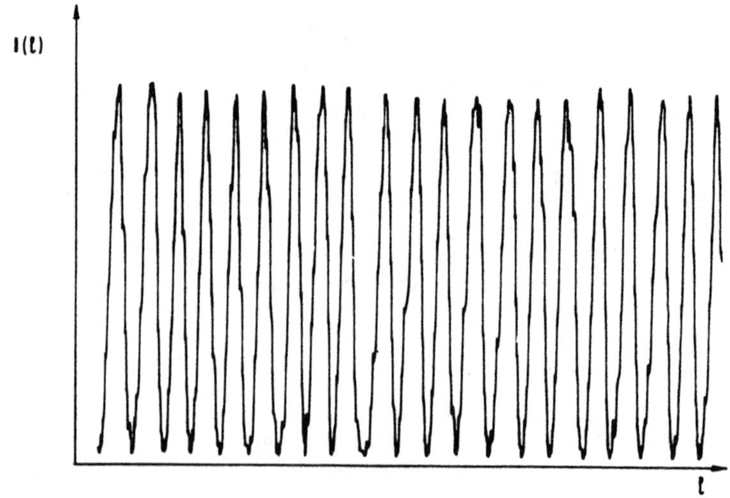

FIG. 6. Helium–neon laser interferogram.

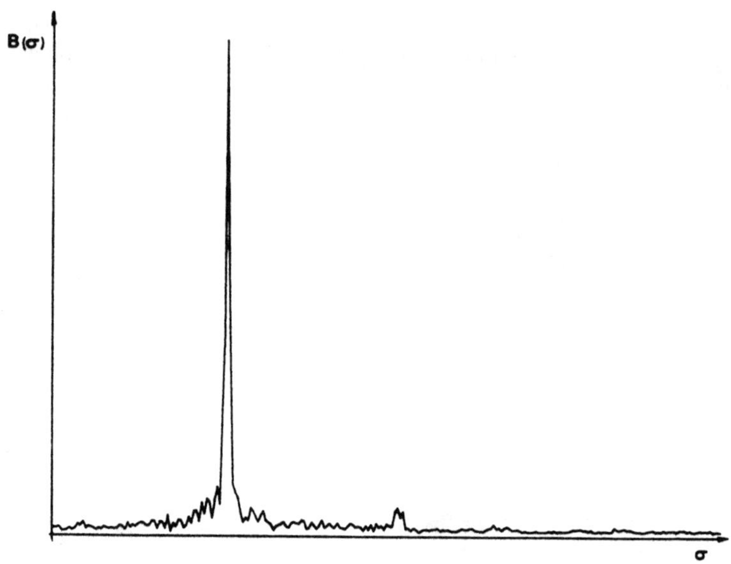

FIG. 7. Helium–neon laser spectrum.

TABLE. 1. Demonstration program.

```
97  REM  ******************************
98  REM  *** FFT LABORATORY PROGRAM ***
99  REM  ******************************
100  CLEAR
110  LOMEM: 16385
120  DIM RE(1024)
130 D$ =  CHR$ (4)
140 U(1) = 1:U(2) = 3:U(3) = 2:U(4) = 6:U(5) = 4:U(6) = 5
150 D(1) = 5:D(2) = 4:D(3) = 6:D(4) = 2:D(5) = 3:D(6) = 1
160  POKE 49337,0
170  POKE 49336,255
180  POKE 49337,4
190  POKE 49333,48
200  POKE 49332,0
210  POKE 49333,52
220  INPUT "SPACING";M
230  INPUT "SAVE ON DISC? Y/N";B$
240  IF B$ = "N" THEN  GOTO 270
250  INPUT "NAME OF INTERFEROGRAM FILE";F1$
260  INPUT "NAME OF SPECTRUM FILE";F2$
267  REM  ****************************
268  REM  *** RECORD INTERFEROGRAM ***
269  REM  ****************************
270  FOR X = 1 TO 1024
280 RE(X) = 0
290  FOR W = 1 TO 10
300 I =  PEEK (49332)
310 RE(X) = RE(X) + I
320  NEXT W
330 RE(X) = RE(X) / 10
340  PRINT X,RE(X)
350  FOR Y = 1 TO M
360  FOR Z = 1 TO 6
370  FOR W = 1 TO 5
380  NEXT W
390  POKE 49336,U(Z)
400  NEXT Z
410  NEXT Y
420  NEXT X
430  PRINT "DATA COLLECTED."
437  REM  ******************************
438  REM  *** SCALE AND PLOT ON VDU ***
439  REM  ******************************
440 Z = 1
450  FOR X = 1 TO 1024
460 A = RE(X)
470  IF A < Z THEN  GOTO 490
480 Z = A
490  NEXT
500 D = Z / 158
510  HGR : HCOLOR= 3
520 X = 1: HPLOT X / 4,RE(X) / D
530  FOR X = 2 TO 1024: HPLOT  TO X / 4,RE(X) / D
540  NEXT X
547  REM  *******************************
548  REM  *** SCREEN DUMP ON PRINTER ***
549  REM  *******************************
550  PRINT D$;"PR£1"
560  CALL  - 16046
570  PRINT D$;"PR£0"
```

TABLE. 2. Demonstration program.

```
577  REM  **********************************
578  REM  *** SAVE INTERFEROGRAM ON DISC ***
579  REM  **********************************
580  IF B$ = "N" THEN  GOTO 670
590  PRINT D$;"OPEN";F1$
600  PRINT D$;"WRITE";F1$
610  PRINT 1024: FOR X = 1 TO 1024: PRINT X: NEXT X
620  FOR X = 1 TO 1024
630  PRINT RE(X): NEXT
640  PRINT D$;"CLOSE";F1$
647  REM  *******************
648  REM  *** FFT PROGRAM ***
649  REM  *******************
650 A = 0:B = 0:F1 = 0:F2 = 0:P1 = 0:P2 = 0:CC = 0
660 SS = 0:I = 0:IS = 0:T = 0:Z = 0:DT = 0
670 L = 0:D = 0:X = 0:Y = 0
680 N0 = 0:N1 = 1:N2 = 2:N3 = 3:N4 = 4
690 N5 = 5:N6 = 6:N7 = 7:N8 = 8:N9 = 9
700 Q = 10
710 P = Q - N1:R = N2 ^ Q - N1:R2 = N2 ^ P:R4 = N2 ^ (P - N1)
720  DIM IM(R),SI(R4)
730 K = 3.142 / R2: FOR X = N0 TO R4
740 SI(X) =  SIN (K * X): NEXT
750  FOR X = N0 TO R
760 IM(X) = N0: NEXT
770  PRINT "IN PLACE FFT AND BIT REVERSAL"
780 TI$ = "000000": GOSUB 1120:TE$ = TI$
790 TI$ = "000000": GOSUB 1330:TE$ = TI$
800 MAX = N0: FOR X = N0 TO R
810 RE(X) =  SQR (RE(X) * RE(X) + IM(X) * IM(X))
820  IF RE(X) > MAX THEN MAX = RE(X)
830  NEXT
837  REM  **************************************
838  REM  *** SCALE AND PLOT SPECTRUM ON VDU ***
839  REM  **************************************
840 Z = 1: FOR X = N0 + 1 TO R:A =  INT (MAX * RE(X))
850  IF A < Z THEN  GOTO 870
860 Z = A
870  NEXT
880 D = Z / 158: HGR : HCOLOR= 3: HPLOT 0,159
890  FOR X = N0 + 1 TO R / 4:A =  INT (MAX * RE(X)):A = A / D
900  HPLOT  TO X,(159 - A): NEXT
907  REM  ******************************
908  REM  *** SCREEN DUMP ON PRINTER ***
909  REM  ******************************
910  PRINT D$;"PR£1"
920  CALL  - 16046
930  PRINT D$;"PR£0"
937  REM  ****************************
938  REM  *** SAVE SPECTRUM ON DISC ***
939  REM  ****************************
950  IF B$ = "N" THEN  GOTO 1030
960 D$ =  CHR$ (4)
970  PRINT D$;"OPEN";F2$
980  PRINT D$;"WRITE";F2$
990  PRINT 256: FOR X = 1 TO 256: PRINT X: NEXT X
1000  FOR X = 1 TO 256:A =  INT (MAX * RE(X))
1010  PRINT A: NEXT X
1020  PRINT D$;"CLOSE";F2$
```

TABLE. 3. Demonstration program.

```
1027  REM  ************************
1028  REM   *** REWIND MICHELSON ***
1029  REM  ************************
1030  FOR X = 1 TO 1024
1040  FOR Y = 1 TO M
1050  FOR Z = 1 TO 6
1060  FOR W = 1 TO 5
1070  NEXT W
1080  POKE 49336,D(Z)
1090  NEXT Z
1100  NEXT Y
1110  NEXT X: PRINT "REWOUND": END
1117  REM  **********************
1118  REM  *** FFT SUBROUTINE ***
1119  REM  **********************
1120  FOR X = N0 TO N9:L(X) = N1: NEXT
1130  FOR X = N0 TO P:L(X) = N2 ^ (Q - X): NEXT
1140  FOR X = N0 TO N9:L1(X) = L(X) - N1: NEXT
1150 X = N0:X1 = N0
1160 X2 = X1
1170  FOR X3 = X2 TO L1(N7) STEP L(N8)
1180  FOR X4 = X3 TO L1(N6) STEP L(N7)
1190  FOR X5 = X4 TO L1(N5) STEP L(N6)
1200  FOR X6 = X5 TO L1(N4) STEP L(N5)
1210  FOR X7 = X6 TO L1(N3) STEP L(N4)
1220  FOR X8 = X7 TO L1(N2) STEP L(N3)
1230  FOR X9 = X8 TO L1(N1) STEP L(N2)
1240  FOR Y = X9 TO L1(N0) STEP L(N1)
1250  IF X >  = Y GOTO 1270
1260 A = RE(X):RE(X) = RE(Y):RE(Y) = A
1270 X = X + N1
1280  NEXT : NEXT : NEXT : NEXT
1290  NEXT : NEXT : NEXT : NEXT
1300 X2 = X2 + L(N9): IF X2 <  = L1(N8) GOTO 1170
1310 X1 = X1 + N1: IF X1 <  = L1(N9) GOTO 1160
1320  RETURN
1330  FOR S = N0 TO P: PRINT "STAGE",S
1340 T = N2 ^ S:T1 = T - 1:IS = N2 * T
1350 D = N2 ^ (P - S):DT = IS * (D - N1):L = N0
1360  FOR Z = N0 TO T1: IF L < R4 GOTO 1380
1370 SS = SI(R2 - L):CC =  - SI(L - R4): GOTO 1390
1380 SS = SI(L):CC = SI(R4 - L)
1390  FOR I = N0 TO DT STEP IS:A = I + Z:B = A + T
1400 F1 = RE(A):P1 = CC * RE(B) - SS * IM(B)
1410 F2 = IM(A):P2 = SS * RE(B) + CC * IM(B)
1420 RE(A) = F1 + P1:IM(A) = F2 + P2
1430 RE(B) = F1 - P1:IM(B) = F2 - P2
1440  NEXT :L = L + D: NEXT : NEXT
1450  RETURN
```

The white light spectrum is heavily modified by the spectral response of the phototransistor and a much improved spectrum would be obtained using a detector such as a PbS cell which has a wider spectral response in this region, the system as it stands however provides a useful introduction to the technique with the possibility of further extensions into the area of absorption spectra.

REFERENCES

1 B. Rogers: *Practical Computing* **91** (December 1980).
2 W. Barbiz: *Practical Computing* **112** (September 1981).
3 D.K. Berkey: *American Journal of Physics* **40**, 267 (1962).
4 J. Davies: *American Journal of Physics* **45**, 166 (1977).

Frequency modulation of laser light

D.E. Evans

Department of Physics, University of Birmingham, P.O. Box 363, Birmingham B15 2TT

Abstract: A Michelson interferometer has one of its mirrors mounted on a piezo-electric crystal. A low-power He–Ne laser is used as the light source, and a photo diode is used to monitor the light intensity in the interference pattern. The motion produced by a time-varying signal causes the interference pattern to fluctuate. Various effects which may be demonstrated with this, including transmission of audio signals, are described.

When a Michelson interferometer is set up normally with monochromatic light, circular fringes are observed in the field of view. As one of the mirrors is moved, the circles are observed to expand (or contract), new fringes forming (or disappearing) in the centre of the pattern each time the path differences changes by λ. It is clear that a photodetector placed so that it records the intensity of the centre of the pattern will record a varying intensity as the fringes move over it.

An alternative explanation of the observed variation is that the light reflected from the moving mirror is Doppler-shifted on reflection, and the variation is simply the beat between the frequency-shifted light and the non-shifted light reflected from the stationary mirror[1]. This experiment demonstrates what occurs when the mirror is moved by an audio frequency signal, which need not be a pure sinusoid. The results can be interpreted either as to-and-fro motion over a fringe or as frequency modulation of light, whichever is more convenient.

The apparatus (Fig. 1) consists of a standard Michelson interferometer, with one mirror mounted on a piezo-electric crystal P. The piezo voltage is derived from a high-voltage amplifier, (outputs of 0–250 V over the audio-range) (Fig. 2a). The input to the amplifier is via a unity-gain op-amp with d.c. offset, so that any audio source may be used. There is no reason why a loudspeaker coil should not be used to drive the mirror, except for space limitations.

The recombined beam falls on a photo diode, which feeds directly into a current-to-voltage convertor (Fig. 2b). The photosignal may be displayed on a CRO or fed into a power amplifier and loudspeaker. Alternatively, it may be digitized and stored in a micro–computer, and then Fourier transformed to display the frequency components. The graphs (Figs 3 and 4) shown are 'screen-dumps' of such data stored in a BBC model B microcomputer.

The first step in the experiment proper is to calibrate the motion of the

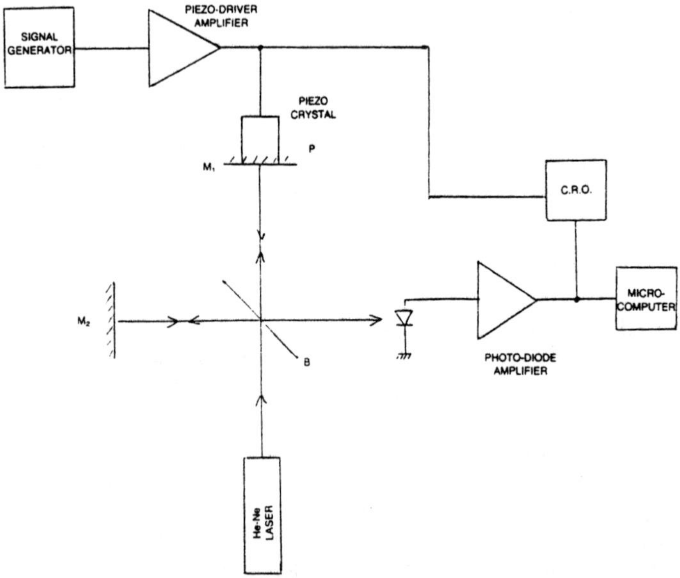

FIG. 1. Block diagram of apparatus.

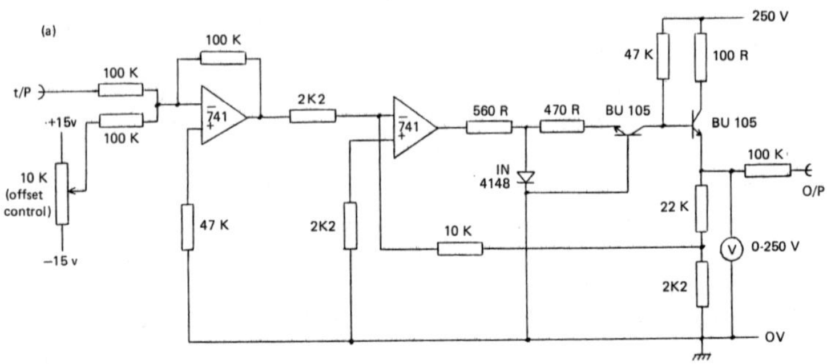

FIG. 2. (a) Schematic circuit diagram of high-voltage amplifier (piezo driver). The transistors are BU105 or equivalent, the op-amps are 741 or equivalent.

(b)

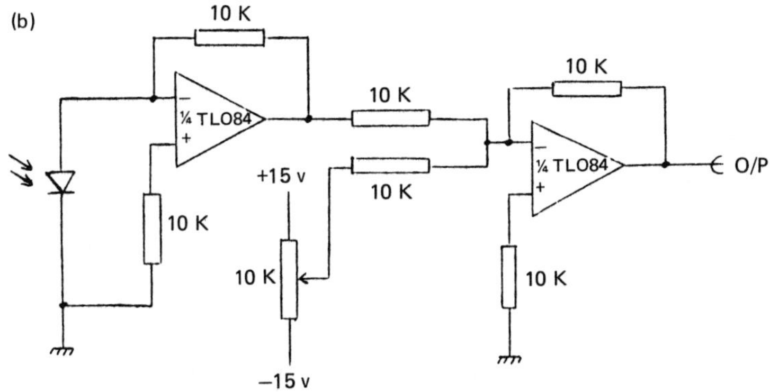

FIG. 2. (b) Schematic circuit diagram of light detector. The second op-amp provides a d.c. offset suitable for driving a DAC such as the ones in the BBC micrcomputer.

piezo-mounted mirror M_1. As the displacement is directly proportional to the voltage applied to the piezo P, monitoring the latter will allow the instantaneous displacement to be deduced. Calibration is achieved by applying a triangle voltage to P, and recording the resultant light intensity I (Fig. 3). The variation of I with time t is easily explained. Starting at the lowest point A on the triangle, one is at an arbitrary point on the fringe pattern, which depends on the path difference between paths BM_1B and BM_2B (Fig. 1). As V increases with t, the mirror moves forward until the minimum light intensity corresponding to full destructive interference is reached (point B). The mirror continues to move forward until the path difference corresponds to full constructive then to interference (C), and destructive interference again (D). Finally one reaches to the peak of the triangle waveform (E). The motion then reverses, and the fringe pattern scanned in the opposite direction, producing the mirror image signal. Knowing the peak-to-peak voltage V_o, the voltage swing V corresponding to a movement of the mirror through a distance $\lambda/2$ is calculated from

$$V = V_o(D - B)/(E - A)$$

and the distance x moved when the voltage is changed by 1 volt is then readily found using

$$x = \lambda/(2V)$$

For our apparatus, x is about 1–2 nmV^{-1}.

The way is now open to do many interesting things. For instance, the peak-to-peak voltage may be adjusted so that the displacement is exactly $\lambda/4$. Scanning from the minimum (B) to the maximum (C) and back will

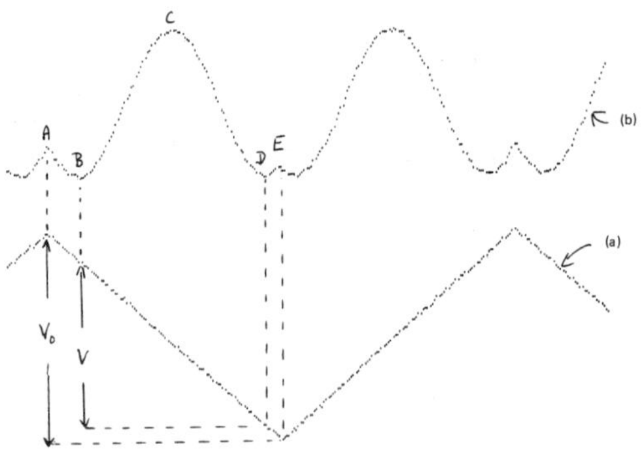

FIG. 3. Voltage calibration: (a) input waveform to piezo driver; (b) corresponding light signals. The points A–E are used as explained in the text.

then produce a fringe pattern which varies with time in exactly the same way as if M_1 had been moved at a steady speed over many fringes (Fig. 4a). The fringes here have a pure cos ωt dependence. But by shifting the d-c offset on the input signal, or by moving M_1 with the microcrometer built into the interferometer, it is possible to start the scan half way up the fringe, and scan over the maximum to the half-way point on the other side, producing a time variation reminiscent of a full-wave rectified signal, with a strong second harmonic content (Fig. 4b). In fact, if a sinusoidal waveform with the same peak-to-peak voltage is used, then the cusps at the turn-round point will be removed, because the mirror moves slowly in this region, and one gets almost perfect second harmonic generation (Fig. 4c).

The sensitivity of the system to the rest position of the mirrors has interesting consequences from the frequency modulation viewpoint. The problem is that the reference or 'local oscillator' signal is derived from the unmodulated carrier, and hence is fully coherent with it. One is in effect using homodyne detection with a phase-sensitive detector! This aspect of the experiment is being written up for publication elsewhere.

Finally, it is not necessary to use a signal generator. The output of a cheap portable radio was connected to the input amplifier and the output was fed into a power amplifier and loudspeaker. The reproduced sound of Radio 1 was better (i.e. less poor!) quality than that on the radio's own loudspeaker, even though the system is not designed to operate at high frequencies (the piezo-crystal has a mechanical resonance at a few hundred Hz). The fact that light was carrying the signal can be demonstrated simply by blocking the modulated beam. The fact that it is f.m. and not an a.m. signal can be demonstrated by blocking the unmodulated light from M_2.

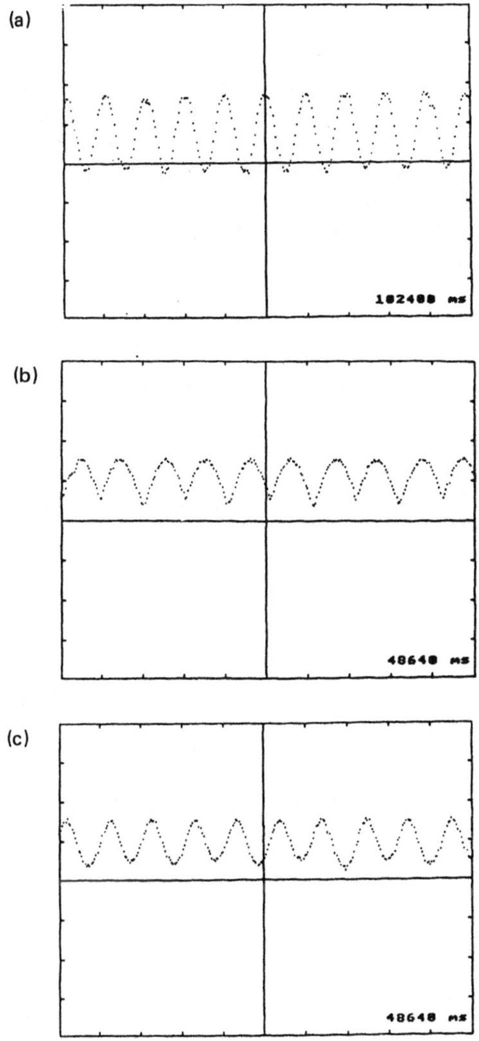

FIG. 4. Screen dumps showing various effects when the peak-to-peak displacement of the piezo crystal is exactly $\lambda/4$.
(a) Triangle wave input, scan starting at minimum of interference pattern (point B on Fig. 3) and finishing at maximum (C), giving pure fundamental sine wave.
(b) Same input as (a), but with scan starting half way up the interference fringe, giving a 'full-wave rectified' signal (time scale $\sim 2\times$(a)).
(c) Sine-wave input, removing the cusps in (b) and producing pure second harmonic.

Acknowledgements

I would like to thank Dr. C.O. Blyth for suggesting and encouraging this work, Dr. N. Berovic for the loan of the piezo crystal, mirro M_2 and the high-voltage amplifier, and Dr. C.E. Gough for the FFT program. Special mention must go to Andrew Jones for some of the ideas in the experiment, especially his work on second harmonic generation and on the audio possibilities. Last but not least, I would like to thank the many PH 270 students who have worked on this experiment, some of whose results appear as Figs 3 and 4.

REFERENCE

1 K.N.R. Taylor, M.B. Allenson, A.R. Piercy: *Physics Education* **5** (1960).

An infrared spectrometer

I.R. KENYON
Physics Department, Birmingham University, P.O. Box 363, Birmingham B15 2TT

Abstract: An old single arm Grubb-Parsons infrared spectrometer has been given a new lease of life by applying modern technology. The Rotational-Vibrational bands of simple diatomic molecules are recorded by students and determinations made of the nuclear separation, spring constant and the population of the angular momentum states.

The study of vibrational-rotational absorption bands of diatomic molecules is obviously interesting at undergraduate level. Using results learnt in first year mechanics about harmonic oscillators and rotating dumbells it is simple to calculate how the frequencies seen in the absorption abnds depend on the spring constant of the force between the atoms and their separation. The pattern of P– and R– branches with nearly equally spaced lines and the absence of the central line illustrates well the selection rule $\Delta J = \pm 1$. Many lines are present (for example with CO or HCl gas) because many rotational levels are populated at room temperature. In fact the modulation of the pattern contains information on this population distribution. Calibration of the instrument is simple because atmospheric absorption gives well marked CO_2 and H_2O bands centred at 4.26 and 2.67 μm respectively.

An old single arm Grubb-Parsons infrared spectrometer has been given a new lease of life in the second year Undergraduate Optics Laboratory after languishing in a cupboard for a number of years. The Nernst glower source was broken and has been replaced by a 10 cm coil of Brightray 26 gauge wire run off 10 volts. This is a much weaker source although cheaper (10 p against several hundred pounds). However, by employing a recently marketed Mullard P1604 pyroelectric detector the signal level (~ mV) achieved is more than adequate. The pyroelectric detector is a ceramic (doped triglycine sulphate) that responds to changes in temperature by changes in its degree of electrical polarization.

Figures 1 and 2 show the opto-mechanical layouts. Radiation from the source is chopped at 16.6 Hz and is directed to the input slit of a monochromator. The 300 lines/mm grating can be driven manually or by a constant speed motor so that the scanning rate in wavelength $d\lambda/dt$ is constant. From the exit slit the output beam passes through the specimen cell and is focussed onto the P1604 detector. After amplification the signal is ~ mV on several volts background. A phase sensitive detector is used to pick out the signal, the reference square wave being provided by an

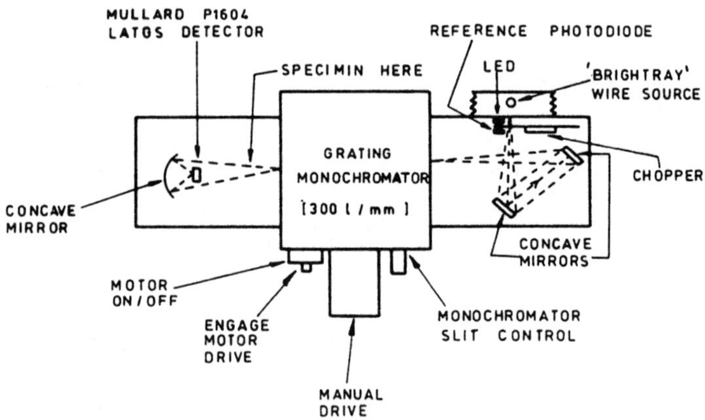

FIG. 1. Opto-mechanical schematic.

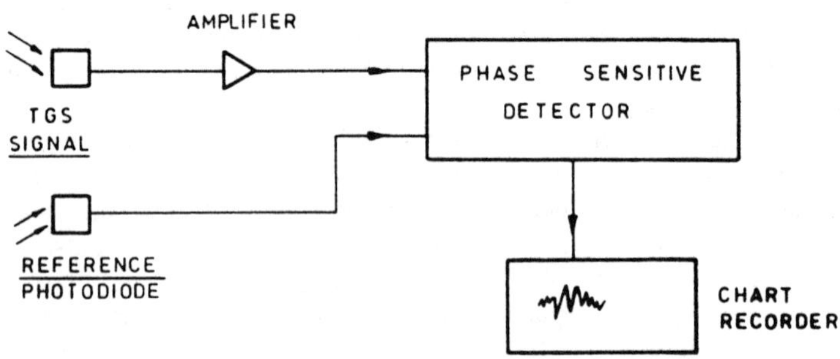

FIG. 2. Electronics schematic.

LED/photodiode pair positioned one on either side of the chopper vane. The phase relative to the signal can easily be adjusted by moving this pair laterally. The output of the PSD is recorded on a chart recorder. It proves convenient in practice to monitor the signal and reference on an oscilloscope. Notice that a low frequency for the chopper is dictated by the detector operation: the temperature of the bulk material has to rise before its polarization changes. The incorporation of the PSD is quite natural and gives the student a good introduction to the device. Figure 3 shows the atmospheric absorption spectrum (no sample) obtained by a pair of students. In the absence of absorption the curve would be smooth and its peak corresponds well to the position expected for the spectrum of a black body at the temperature of the source. Various CO_2 and H_2O absorption troughs are seen. There is also a trough at ~3.4 μm which was present only whilst the building containing the apparatus was being repainted. This trough appears to come from absorption by CH bonds in the paint solvent. A scan as shown in Fig. 3 requires about 80 minutes and serves to calibrate the instrument. Figure 4 shows the absorption spectra obtained with an HCl gas cell in place. The cell is 10 cm long, has NaCl windows and was filled at one atmosphere pressure.

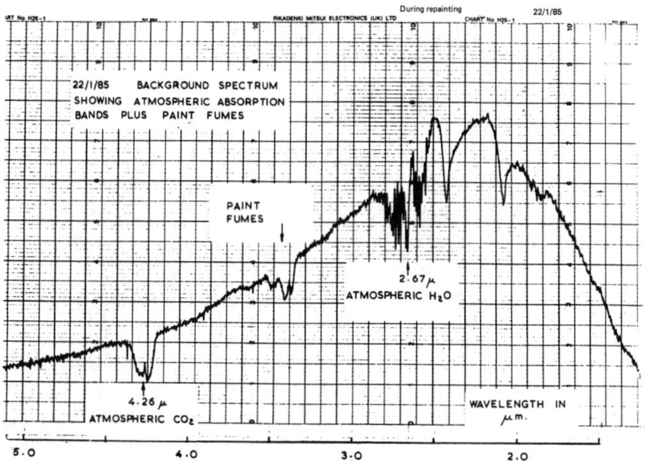

FIG. 3. The emission spectrum of the source with absorption bands due to the atmosphere plus paint fumes in the laboratory.

The chapter on Molecular Spectra in the book on *Quantum Physics* by Eisberg and Renick usefully summarizes much of the basic physics. Measurements of the spacing of the lines in the band spectra for diatomic gases yields the quantity:

$$D(1/\lambda) = h/4\pi^2 Ic$$

where I is the moment of inertia of the molecule about an axis perpendicular to the molecular axis. The moment of inertia is given by:

$$I = \mu r^2$$

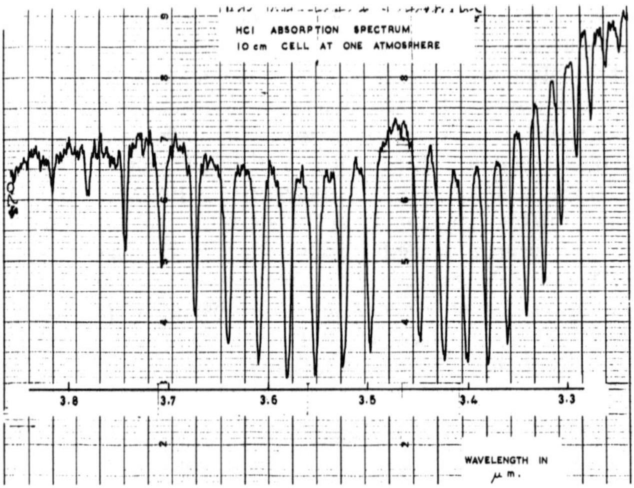

FIG. 4. The HCl absorption spectrum.

where μ is the reduced mass and r the separation of the nuclei. If λ is the wavelength of the centre of the vibrational rotational band the spring constant, f, of the force between the atoms in the molecule is given by:

$$f = 4\pi c^2\mu/\lambda^2$$

Finally the depth of each absorption line depends on the population of the parent rotation level. These levels are populated according to the Boltzmann distribution taking account of the degeneracy of each level:

$$n_J = n_0\,(2J + 1)\exp\left(-(E_J - E_0)/kT\right)$$

Here $E_J - E_0 = J(J + 1)h^2/8\pi^2 I$

It is possible to check the expected distribution against that observed from the modulation depth.

The project takes 7 hours laboratory time and around 1–2 hours of work outside the laboratory.

Acknowledgements

I would like to offer my thanks to Mr. Steven Wotton, a second year undergraduate, who spent several weeks last summer bringing the device into operation.

I would also like to thank members of the staff who helped in various ways and at various times: Dr. John Glen, Dr. John Griffiths and Dr. Chris van Zyl

REFERENCES

1 R. Eisberg and R. Resnick: *Quantum Physics*. John Wiley, New York (1974).
2 C.N. Bamwell: *Fundamentals of Molecular Spectroscopy*. McGraw-Hill Publishing (1966).

Coherence length

C.H. PANTER
School of Physics, Newcastle upon Tyne Polytechnic, Ellison Building, Ellison Place, Newcastle upon Tyne NE1 8ST

Abstract: An experiment to correlate the coherence length of radiation with its relative bandwidth is described. The experiment is intended to give the student an appreciation of the significance of the bandwidth of the radiation in interferometry. Examination of the results shows that the wave trains appear to be damped.

Most, if not all, undergraduates on a physics-based course are familiar with the Michelson interferometer. In looking for an experiment that would be a little more demanding than a straightforward wavelength measurement, consideration of the equipment immediately available in the laboratory led to the conclusion that coherence length measurement was ideal. If it was measured for different bandwidths the results could then be compared with what one can predict using alternative simple models for coherence phenomena. It was felt that this would be an interesting change from the usual routine of undergraduate laboratory work to give the student the task of deciding which of the two models gives the best fit to the results, thus illustrating the relationship between theory and practice in physics.

As can be seen from Fig. 1, the apparatus is very simple. The source needs to be as intense as possible due to the loss of light at small slit widths

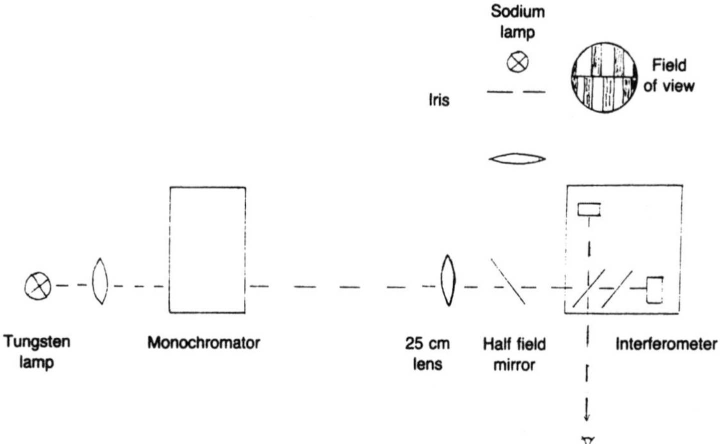

FIG. 1. Block diagram of apparatus.

on the Hilger and Watts monochromator. We used a 12 V tungsten halide lamp followed by an f/1, 5 cm lens. In order to obtain a satisfactory field in the interferometer, it has been found that the incident light should be in the form of a converging beam of small cone angle. For this reason, it is necessary to place a 25 cm lens as in Fig. 1 at about 35 cm from the exit slit of the monochromator. This is the most critical component in the equipment. The interferometer used is a fairly early Beck model which had rather coarse control of the mirror movement. This was improved by attaching a 75 cm aluminium disc to the thimble of the micrometer. Since at maximum bandwidth, the mirror movement was only 0.02 mm, it was decided to use a second set of interference fringes, from the sodium source, for the measurement of the mirror displacement.

In carrying out the experiment, the student first calibrates the slit width, in terms of bandwidth, by means of a cadmium lamp, then proceeds to set up the apparatus for the main part of the experiment.

Due to the use of the slits, the geometry of the apparatus is quite critical, but once the student is aware of this, it usually presents little difficulty. The interferometer must of course be set up for zero path difference. This is most conveniently done by using sodium light interference fringes. Circular fringes are first produced, then the mirrors are inclined thus offsetting the centre of the fringes. The moving mirror is now displaced so that the centre of the fringes is on the opposite side of the field of view to that previous. Midway between these positions is a position where the fringes are most distinct and straight — corresponding to the zero path difference position. This can then easily be checked using a convenient white light source.

Measurements of coherence length are made for various bandwidths at a few selected wavelengths. Two types of wave train may be considered.

(a) Totally undamped with a duration λ and

(b) Subject to exponential damping with decay constant λ from the moment of initiation. It can be shown that from these models one can specify a coherence length l given by $l = 2\lambda_o^2/\Delta\lambda$ and $l = \lambda_o^2/\pi\Delta\lambda$ respectively where λ_o is the mean wavelength and $\Delta\lambda$ is the total bandwidth in both cases.

Figure 2 shows typical results of a plot of l versus $\lambda_o^2/\Delta\lambda$, from which it will be observed that the latter model is the better. Typically, the experiment can be performed in about 5 to 6 hours and provides a change from the normal routine experiments.

Since the exhibition, some consideration has been given to improvements. The use of a 10–15 cm lens in front of the interferometer with a corresponding re-positioning of the 25 cm lens makes it much easier to

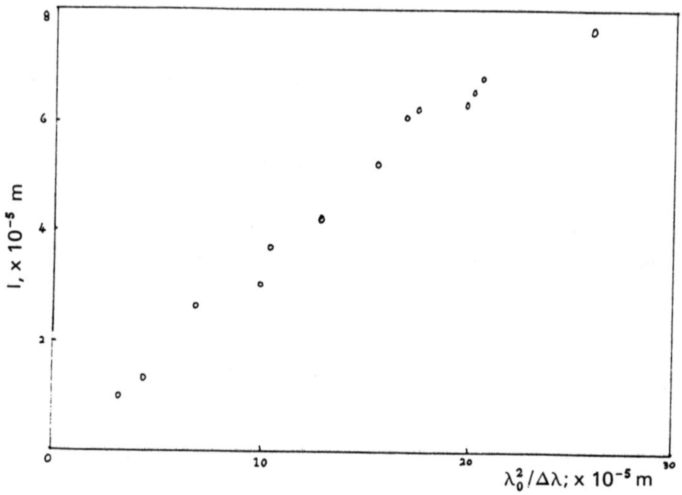

FIG. 2. Typical plot of coherence length l against mean wavelength/relative bandwidth.

observe the fringe movement relative to the marker. Since the fringe movement is still rather rapid we have since tried a Hilger and Watts model. This is a much nicer instrument with respect to control of the fringe movement but is unfortunately more difficult to set up in the first instance.

REFERENCES

1 A.C.S. Van Heel: *Advanced Optical Techniques*. North Holland (1967).

The visibility of sodium fringes

A.W. Williams
Department of Physics, University College of Swansea, Singleton Park, Swansea SA2 8PP

Introduction

The visibility of interference fringes is defined by

$$V = (I_{max} - I_{min})/(I_{max} + I_{min})$$

where I_{max} and I_{min} are the maximum and minimum intensities of the fringe pattern. The visibility will decrease with increasing path differences between the two interfering wavefronts and at a path difference defined as the coherence length of the light source, the visibility tends to zero. For a single spectral line source the visibility falls off smoothly with path difference (Fig. 1(a)), while for a source consisting of more than one spectral line the visibility curve will show considerable structure (Fig. 1(b)). A sodium lamp radiates most of its light energy in the lines of a doublet having almost equal intensities and will give a visibility curve with regular subsidiary minima in which the visibility is almost zero (Fig. 1(c)).

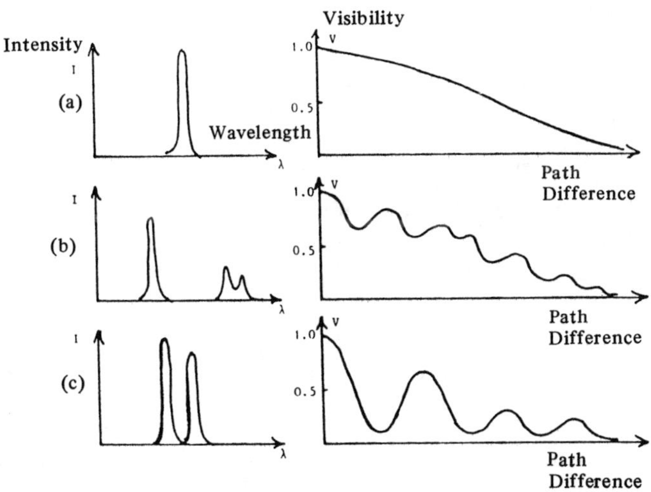

FIG. 1. Visibility as a function of path difference for various spectral distributions: (a) Monochromatic source; (b) A source with many lines; (c) A doublet source.

Apparatus and experimental details

A Michelson interferometer, focusing lens and photomultiplier are mounted on an optical bench. The interferometer is adjusted to give circular fringes with a sodium lamp and the fringe pattern is focussed on to a plate covering the photocathode of the photomultiplier. The plate has a pin-hole aperture which is smaller than the focussed fringe width. The output from the photomultiplier is displayed on a chart recorder and the path difference is slowly and uniformly increased by turning the graduated drum geared to the moveable mirror of the interferometer by means of a rubber belt driven by an electric motor. The chart recorder amplifier setting, photomultiplier voltage and chart speed are all adjusted so that the zero level, maximum and minimum signals and the distance between the maxima have convenient values. The system is allowed to run until a number of maxima and minima in the visibility curve have been observed. This will usually take more than an hour. During this time the chart and mirror speed should be measured so that distance on the chart may be calibrated in terms of path difference.

Theory

Coherence length

The coherence length is obtained directly from the visibility curve and is the path difference at which the visibility becomes zero.

Wavelength separation of the doublet

At the first maximum in visibility:

For a line at wavelength λ_1 $$2nt_1 \cos\theta = p_1\lambda_1 \quad (1)$$

where n = refractive index of air, θ = angle of fringes, t_1 = position of moveable mirror, p_1 = fringe number.

For a line at wavelength λ_2 $$2n_1t_1 \cos\theta = p_2\lambda_2 \quad (2)$$

Hence $$p_1\lambda_1 = p_2\lambda_2 \quad (3)$$

At the next maximum in visibility, let t_2 be the position of the mirror.

For λ_1 $$2nt_2 \cos\theta = (p_1 + p)\lambda_1 \quad (4)$$

For λ_2 $$2nt_2 \cos\theta = (p_2 + p + 1)\lambda_2 \quad (5)$$

Again $$(p_1 + p)\lambda_1 = (p_2 + p + 1)\lambda_2 \quad (6)$$

Substituting for $p_1\lambda_1$ from (3) $$(\lambda_1 - \lambda_2) = \lambda_2/p \quad (7)$$

Subtracting (1) from (4) $\quad 2n(t_2 - t_1) \cos \theta = p\lambda_1$

For small θ and $n = 1$ $\quad 2(t_2 - t_1) = p\lambda_1$

Substituting for p from (7) $\quad (\lambda_1 - \lambda_2) = \lambda_1\lambda_2/[2(t_2 - t_1)]$

Then $\Delta\lambda = (\lambda_1 - \lambda_2) = \lambda^2/[2(t_2 - t_1)]$

where λ is the average wavelength.

$(t_2 - t_1)$ may be measured directly from the visibility curves.

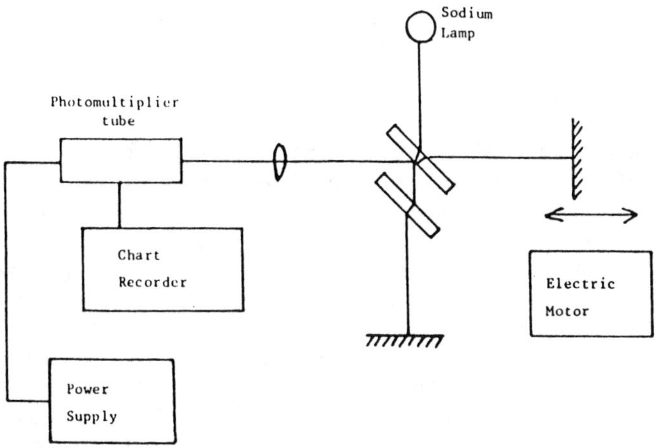

FIG. 2. Block diagram of apparatus.

Average wavelength

At maximum visibility the fringes satisfy

$$2nt_1 \cos \theta = p_1\lambda$$

$$\text{i.e.} \quad 2t_1 = p_1\lambda$$

For the next maximum

$$2t_2 = p_2\lambda$$

$$(t_2 - t_1) = (p_2 - p_1)\lambda/2$$

By counting the number of fringes $(p_2 - p_1)$ between the maxima and knowing $(t_2 - t_1)$, the average wavelength is found.

Properties and applications of a phase sensitive detector

D.P. Woodruff
Physics Department, University of Warwick, Coventry CV4 7AL

Abstract: The phase sensitive detector is widely used in signal recovery in modern physical measurements and this experiment aims to familiarise students with the characteristics of the device as a 'black box'. In particular students investigate the amplitude–phase relationship for a single harmonic input, the bandpass characteristics as a function of time constant, and demonstrate the possibility of using the device as a frequency spectrum analyser. The phase sensitive detector is also used in a specific application, the investigation of contact potential differences between metal surfaces using a vibrating capacitor method.

Introduction

Whilst it is generally regarded as important nowadays that a physics student should have some familiarity with electronics at the level of components and circuit design, it is probably even more important that he learns how to make the best use of electronic instruments without a detailed knowledge of the internal circuitry. An increasingly common problem in physical measurements is the extraction of low level signals from noise, and it is often solved by the use of a phase sensitive detector (PSD), otherwise known as a lock-in amplifier. Typically, an application originally involves the measurement of a d.c. current or voltage at a low level which is susceptible to both noise and drift. This can usually be converted into an a.c. measurement by modulating the source of the signal; for example, in optical measurements, a rotating optical 'chopper' can be used. The a.c. signal can then be separated from much of the noise and from the drift by filtering, the signal now having a known frequency. Stretching this approach by using narrower and narrower band-pass filters ultimately leads to problems, however, associated with drift in the frequency of either the modulating source or the filter (or both). The phase sensitive detector overcomes this by providing a filter of extremely narrow band pass which is locked in phase (and thus frequency) to that of the modulation source which is provided as a 'reference' signal to the detector.

This experiment, undertaken by second year physics students at Warwick, is designed to allow a student to appreciate the basic operation of the PSD and test one simple application. It falls into two main parts; firstly an investigation of the characteristics of the PSD to inputs derived from standard audio frequency function generators and secondly setting up and using a method of measuring contact potential differences between surfaces using a vibrating capacitor method both without, and with, the PSD as a detector.

Investigating the PSD and its operation

The PSD is explained to the student as having the same function as a full wave rectifier but with the switching of the sign of the *signal* being controlled by the sign of the *reference* input. This basic information allows the student to account for the results of this part of the experiment in a quantitative fashion and thus appreciate the mode of operation of the instrument as a 'black box'. The instrument provided is rather simple and was designed and constructed in our departmental electronics workshop. A detailed circuit is available on request. The main feature which distinguishes the instrument from commercial ones, apart from its cost (less than £100), is that several extra outputs are provided so that the student can inspect the signal at different points in its processing. The quality of performance is no doubt inferior to a commercial instrument in many ways, but it is entirely adequate for the present purposes. Figure 1 shows a block diagram of the complete instrument. Note in particular that the phase-shifted reference signal can be inspected, as can the switched and amplified signal after the main PSD stage but before the low pass filter which generates the usual d.c. output as displayed on a panel meter on the instrument. The main requirements of this part of the experiment are as follows:

(a) Using a single harmonic input, which is also fed into the reference channel, investigate the dependence of the d.c. output signal on the phase difference between the signal and reference inputs to the PSD section.

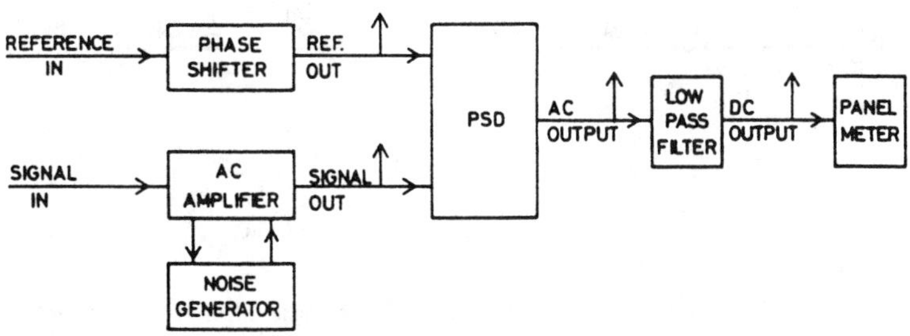

FIG. 1. Block diagram of the complete PSD instrument used in this experiment.

This phase difference can be obtained by displaying both the input and the phase-shifted reference on a two-beam oscilloscope. The phase shifter is uncalibrated, and indeed rather non-linear, but a reasonable set of phase differences between 0° and 180° can be achieved. Figure 2 shows typical data from such a measurement. Students are encouraged to determine the functional relationship theoretically and to test this prediction. Evidently, for a signal of peak-to-peak, amplitude V_o and a reference phase difference of ϕ, the smoothed, chopped output will be

$$\dot{V} = (V_o/2\pi)\{\int_{\phi}^{\pi+\phi} \sin\theta \, d\theta - \int_{\pi+\phi}^{2\pi+\phi} \sin\theta \, d\theta\}$$
$$= (2V_o/\pi) \cos\phi$$

i.e. the functional dependence is simply on the cosine of the phase difference. In fact the experimental data do show this general form but clearly show a zero output cross-over slightly displaced from the predicted phase difference of 90°. This can be ascribed to some additional phase shift in reactive components seen by the oscilloscope and not by the PSD (or vice versa). Plotting the normalised output against corrected phase gives a good straight line to confirm the prediction.

(b) Using two separate oscillators for the reference and signal inputs, both providing harmonic signals, investigate the band-pass characteristics of the complete instrument using different time constants in the band-pass filter. Because it is clearly not possible to set both oscillators at exactly the same frequency, the panel meter actually shows a.c. oscillations in this experiment with an amplitude which falls rapidly as the two input frequen-

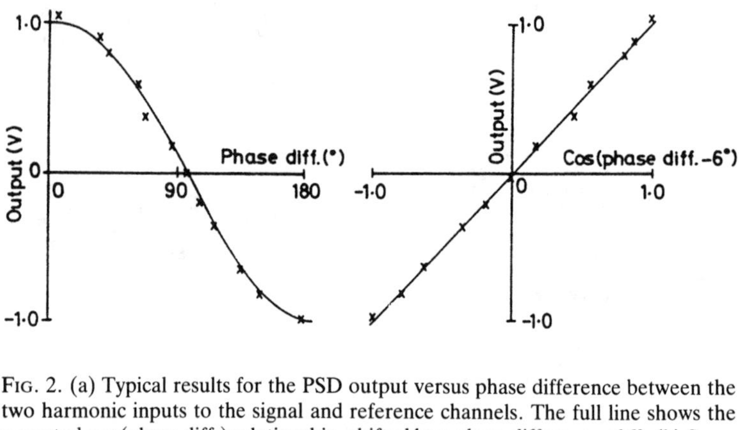

FIG. 2. (a) Typical results for the PSD output versus phase difference between the two harmonic inputs to the signal and reference channels. The full line shows the expected cos (phase diff.) relationship, shifted by a phase difference of 6°. (b) Same data plotted against the cosine of the 'corrected' phase difference.

cies are shifted apart. It is this a.c. amplitude which is plotted in the experiment (e.g. Fig. 3). The viability of this part of the experiment is clearly dependent on the stability of the two function generators used. In fact, at low frequencies, the quite standard commercial oscillators used seem entirely adequate. However, the calibration of the dials on these sources is not perfect and as we are concerned with quite small frequency changes it is helpful to measure the frequencies with a digital frequency meter. Figure 3 shows typical data for three output time constants of 0.1, 1.0 and 10.0 seconds using a nominal signal frequency of 200 Hz. While the full width at half maximum for the two longer time constants (2.5 Hz and 0.23 Hz) is generally compatible with expectation, the lowest time constant gives a bandwidth (3.0 Hz) which is too narrow indicating that the time constant of the panel meter itself is much greater than 0.1 seconds. Removing the smoothing completely leads to data identical to those for the 0.1 sec time constant, confirming this view.

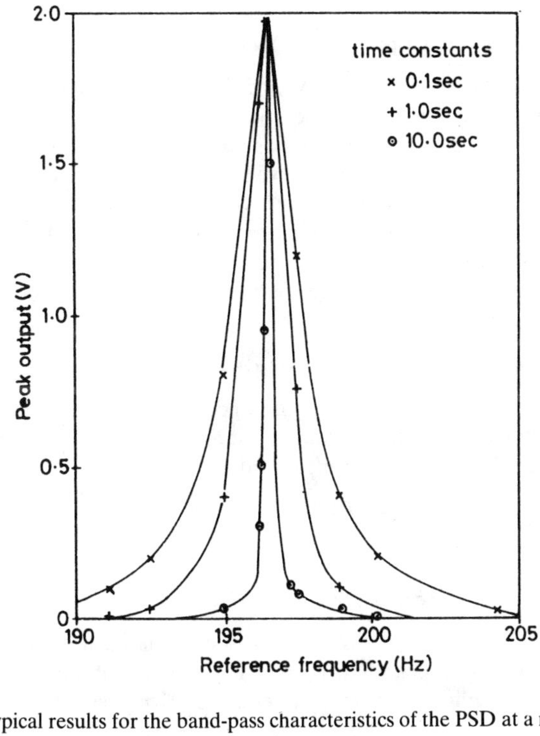

FIG. 3. Typical results for the band-pass characteristics of the PSD at a nominal 200 Hz measured at three different output time constants.

(c) As the primary purpose of a PSD is ultimately to extract signal from noise, a simple simulation is made in which experiment (a) is repeated but with the signal first attenuated (through an external resistor chain), and then reamplified after the addition of noise from a noise generator in the amplifier stage of the PSD. Displaying this amplified signal on an oscilloscope shows that the true signal is barely detectable but an output-phase diagram essentially identical to that in Fig. 2 is obtained.

(d) Using two separate oscillators as in part (b), but with the signal input set to a square waveform, the capability of the PSD to function as a spectrum analyser can be demonstrated. Significant deflections on the panel meter are only found when the reference frequency is essentially identical to that of the signal, or very close to an odd harmonic of the frequency. The relative amplitudes of deflection at these harmonics can be compared (quite favourably) with the predictions of a Fourier series synthesis of the square wave which the students are again encouraged to perform.

Measurement of contact potential differences

If two metal surfaces of different work function are brought together, the inequivalence of the two Fermi levels will cause electrons to flow from one to the other until sufficient charge has built up to level the Fermi levels when the difference in the vacuum levels outside the surfaces (the contact potential diference) will equal the difference in the work functions. This difference can be measured by a vibrating capacitor method. Suppose the two surfaces form a parallel plate capacitor of area A and spacing d. By applying a vibration to one plate we have $d = d_o + d_1 \sin \omega t$ and if we apply a balancing d.c. voltage V_B, the charge on the capacitor is given by

$$q = C(V_B + V_{12}) = (A/\epsilon_o d)(V_B + V_{12})$$

where V_{12} is the contact potential difference. If an amplifier measures the voltage across a resistance R in the circuit associated with the change in q on vibration, then the voltage measured is

$$V = R\mathrm{d}q/\mathrm{d}t \; (= iR)$$

and so

$$V = -(AR/\epsilon_o d^2)(V_B + V_{12})d_1\omega \cos \omega t$$

Evidently if $d_1 << d_o$, the voltage V is essentially sinusoidal. Even if this is not so, V vanishes when $V_B = -V_{12}$: i.e. one obtains a null signal when the balancing d.c. voltage is equal and opposite to the contact potential.

Students set up a system based on this principle and are able to check the null signal both on an oscilloscope and, using the capacitor driving frequency as a reference source, on the PSD. They have a variety of samples (brass, steel, aluminium) and can show the substantial variation in contact potential difference as a function of position on a sample and surface cleanliness due to the marked influence of contamination on this surface property. The absolute values of the contact potential differences measured are of little value because of these contamination problems and measurements of clean surfaces must be performed in ultra-high vacuum. Such an extension is clearly impractical for the present purposes due to the limitations of both cost and time. Typical measurements show variations of contact potentials of $\sim\pm 1$ V or more relative to the aluminium alloy vibrating probe. In principle this reference probe should be coated with an unreactive species such as gold; in practice the fact that the absolute values have little significance renders this subtlety redundant.

The basic circuit diagram for the vibrating capacitor detection system is shown in Fig. 4. The important feature is that a very high value resistance ($10^9 \Omega$) is needed to convert the detected current into a voltage in order that no significant drain is imposed on the system. This also necessitates the use of a high input impedance amplifier stage. The batteries used to provide the voltage rails of the amplifier also provide the balancing offset voltage V_B applied to the probe.

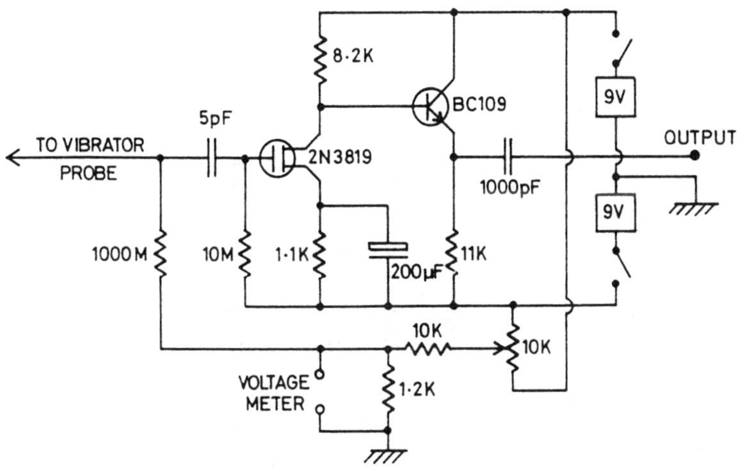

Fig. 4. Circuit diagram for the detection system used in the vibrating capacitor experiment.

Equipment requirements

Most of the equipment used in this experiment is standard laboratory instrumentation, notably a two-beam oscilloscope (sensitivity down to ~10 mV/cm), two low frequency ($\leqslant$ kHz) oscillators with low output impedance to drive the vibrator, and a frequency meter to determine accurate frequencies in the pass band measurements. The offset voltage V_B can be measured using the oscilloscope or with a standard panel meter or multimeter. The only special components are the PSD instrument itself, the vibrating capacitor system amplifier (both built in our electronic workshops), and the low frequency vibrator (Model 101: Ling Dynamic Systems Ltd., Baldock Road, Royston, Herts, SG8 5QB). The electronic component costs were dominated by cases, meters, switches and the $10^9\Omega$ resistor but the total cost of these special components, including the vibrator, are under £200.

Three practical investigations chosen from an undergraduate phenomena laboratory

M.F. Thomas, J.N. Jackson and M.A. Houlden
Department of Physics, University of Liverpool, Liverpool L69 3BX

Abstract: Three investigations are described which are examples chosen from a laboratory that consists of short (3 hour) explorations of a variety of phenomena. For each of the investigations an outline of the aims and procedures are given followed by the instructions and questions just as the student would meet them.

Introduction

The following practical exercises form part of an undergraduate laboratory whose aim is to allow the student to investigate a number of phenomena in a largely qualitative but structured fashion. This laboratory, which complements conventional laboratory work is used by second year undergraduates whose course is composed of ten of the currently available exercises that are listed below. Each exercise is designed to be completed in a single 3-hour session including all writing up and marking. The aim is that the student is guided through the investigation of the phenomenon, prompted to link observations and measurements with physical ideas and is asked questions to test understanding. No experimental write up or long calculations are required.

The current list of investigations in this laboratory are on fluorescence and laser action, ferromagnetic materials, diffraction of electrons, the optical spectrum of hydrogen, the motion of electrons in a magnetic field, the Seebeck and Peltier effects, the design of antennae, crystallography, the thermal population of discrete energy levels, the Faraday effect, the energy gap in silicon and germanium, Stefan's law, nuclear magnetic resonance, electromagnetic waves in a waveguide, the Hall effect, transmission lines, absorption spectra, the Franck Hertz experiment.

In order to best convey the flavour of these investigations a short outline of each is given followed by the written guide and questions just as the student meets them.

A short description of this laboratory and a survey of the response of the first group of students who took this course has been reported[1].

Investigation 1. Optical fluorescence and laser action

This exercise aims to link the ideas of fluorescence and phosphorescence with the concept of metastable states and the role such metastable states play in the action of a laser.

A darkened box has two chambers connected by an opaque shutter. In one chamber a bulb emits ultraviolet radiation, in the other chamber fluorescent samples (activated ZnS powder, fluorescein and activated perspex) are placed. The fluorescence from the samples is viewed through a glass window. The colours of the fluorescence are observed by eye. A direct vision spectrometer is used to view a fluorescent light to identify the discrete lines in the spectrum from the mercury vapour and the continuous fluorescence spectrum.

The shutter between the ultraviolet source and the samples is used to examine the materials for phosphorescence — the persistence of light emission after the exciting ultraviolet radiation has been cut off. The ZnS (Cu) material exhibits this. Questions prompt the concept of metastable states. An outline of the action of the 0.5 mW He–Ne laser is given stressing the role of the metastable state. The cover of this laser has been replaced by a perspex top to make the discharge tube visible. The spectrum of the light from the side of the tube is observed with a Hilger constant deviation spectrometer and the wavelengths of the principal lines measured. The spectrometer is then moved to measure the spectrum of the beam, from behind a ground glass screen. The single line spectrum of the beam is compared with the many lines observed from the side of the tube and the student is asked to explain this.

Student guide

Certain materials can be excited to give off visible light. The excitation process can be bombardment by electrons, heating or, as in this fluorescence study, by irradiation with ultra violet light.

An outline of the process is shown in the diagram below.

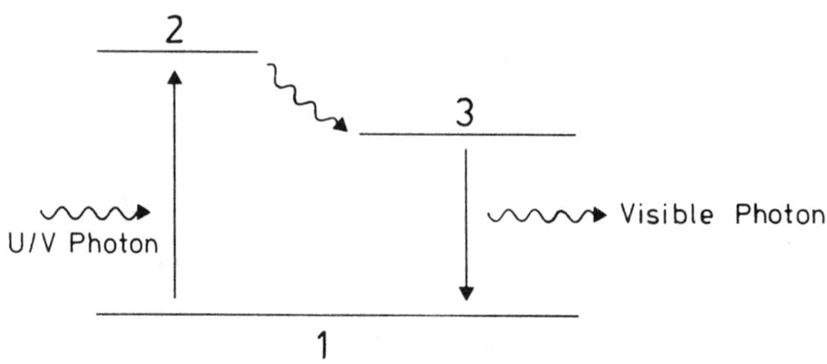

The fluorescent material contains some atom — or group of atoms — that have energy levels as in the diagram above. The process occurs in three steps.

Step 1. A photon (quantum packet of energy $E = hc/\lambda$ is absorbed where, in our case, λ is a wavelength in the ultra violet ($1 \rightarrow 2$).
h = Planck's constant = 6.63×10^{-34} Js
c = velocity of light = 3×10^{8} ms^{-1}
λ = wavelength — for u/v ~200 nm

Step 2 The atom(s) lose energy either by heating the surroundings or by emitting a low energy photon ($2 \rightarrow 3$)

Step 3 Excited state (3) emits a photon of energy $E = hc/\lambda'$ where λ' is a wavelength corresponding to visible light and thus returns to the ground state (1).

Our investigations are designed to fill in some details of this outline.

Simple observation of fluorescence

Put the ZnS phosphors (powders in slide mounts) in the viewing compartment of the green box and view when bathed in the u/v light. (Put on top covers to exclude daylight.) Repeat with the activated perspex and the solution of fluorescein. See Question Sheet Q1.

Information on the light emitting centre can be gained from the spectrum of the fluorescent light given off by the above materials.

If the light emitting levels belong to a set of effectively isolated atoms (atoms not strongly influenced by their surrounding atoms) then the light is emitted at discrete wavelengths.

Alternatively if the light emitting levels belong to an atom or group of atoms that interact strongly with their surrounding atoms — then the levels broaden into bands and the emitted radiation (light) has a continuous spread of wavelengths.

You can observe this by using the direct vision spectrometer to view (i) a discharge tube light where the light comes from gaseous atoms and (ii) a heated wire bulb where the light comes from the strongly interacting atoms of the wire.

The fluorescence from our materials is not sufficiently intense for us to observe its spectrum with our simple apparatus. However in the diagram below the spectra A, B and C correspond to three of the four ZnS phosphors you have viewed. See Question Sheet Q2.

Phosphorescence

Phosphorescent is the name applied to materials whose fluorescence continues after the excitation has ceased.

Using the wooden shutter between the lamp and specimen compart-

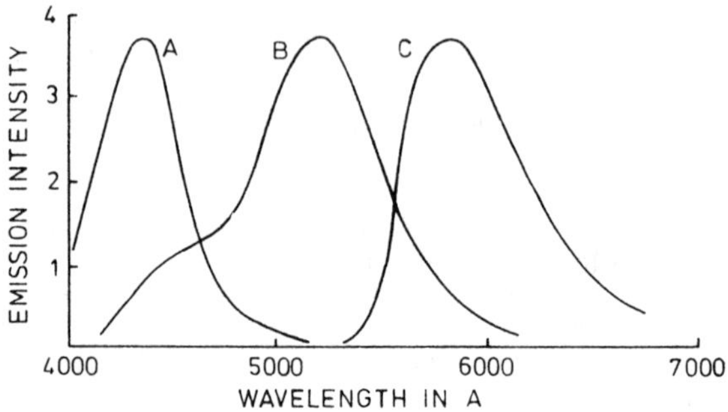

ments first irradiate the ZnS phosphors and then cut off the u/v radiation. Observe if any of the phosphors are phosphorescent. See Question Sheet Q3.

Action of a He–Ne laser

The action of the He–Ne laser is illustrated in the diagram below.

Helium and neon atoms (black dots) constitute the active medium of the gas laser. At the start both are in the ground state (a). Electron bombardment pumps helium to a higher energy level (b). When helium and neon atoms collide, the helium loses its energy to the neon which is raised to one of four distinct energy levels (c). When stimulated by an outside photon, the neon contributes a photon (top wavy arrow) to the laser beam and falls to one of 10 energy levels. The neon then reverts to

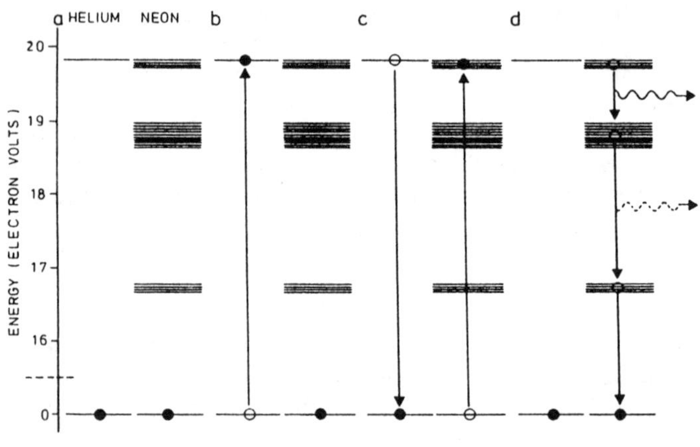

the ground state in steps. Photons emitted in the latter stages of the deexcitation of the neon atoms do not contribute to the laser beam. Thus the laser beam consists of these stimulated transitions.

Observation of the spectra of the laser tube and the laser beam

Using the spectrometer observe the spectrum of the glowing tube of the He–Ne gas laser by setting up the laser and spectrometer as in the diagram. The wavelength of the light radiated can be measured on the calibrated drum of the spectrometer. See Question Sheet Q4.

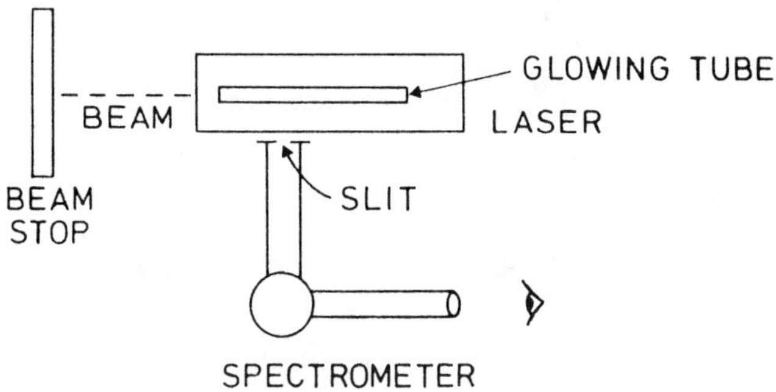

Exercise 4

Consult a demonstrator who will set up the final observation with you. You examine the spectrum of the laser beam observed through the ground glass screen. See Question Sheet Q5.

Further reading

The library contains a number of books on fluorescent (or luminescent) materials and on lasers. In particular, 'Luminescent materials — Garlick' and 'Lasers and Light (reading from *Scientific American*) Schawlow', are not too technical.

Question Sheet

Q1 List the principal colour of fluorescent light from
ZnS (Ag)
ZnS (Cu)
ZnS (Mn)
Activated perspex
Solution of fluorescein

Q2 (i) A fluorescent tube light contains mercury vapour and a fluorescent substance coated on the inside of the tube. View the spectrum of the fluorescent tube with the direct vision spectrometer. Describe the spectrum. State the type of atoms, e.g. Hg vapour atoms or fluorescent material atoms that give rise to the observed spectrum.
(ii) Identify the phosphors;
A
B
C
(iii) Sketch the spectrum you would expect for the phosphor that fluoresces white.

Q3 (i) List any ZnS phosphors that show phosphorescence.
(ii) What is the time (approx) before the phosphorescence fades away?
(iii) Using the level diagram give a very brief explanation of phosphorescence.

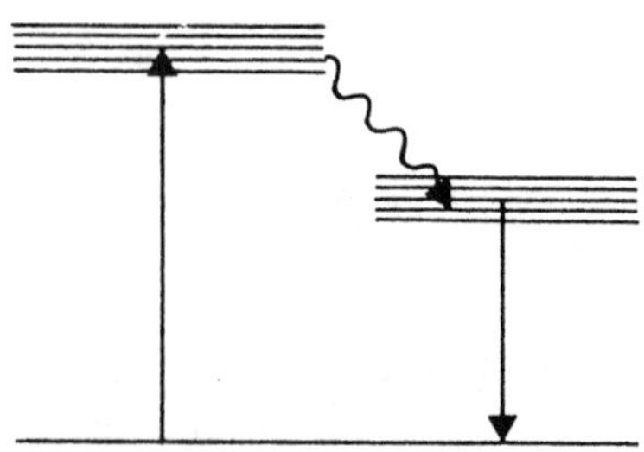

Q4 List the wavelengths and colours emitted from the glowing tube.
λ (nm) Colour

Q5 Measure the wavelength of the laser beam in nm.
λ(Beam) =

Q6 Why is the spectrum of the laser beam a single line?

Ferromagnetic materials

This exercise seeks to link macroscopic properties of ferromagnetic materials such as magnetic hardness with the concept of magnetic domains through consideration of the hysteresis cycle. Magnetic domains are observed in a crystal of yttrium iron garnet (YIG) when viewed by a microscope between nearly crossed polaroids. A current through a coil wound around the crystal mount sets up a magnetic field collinear with the direction of magnetisation and allows the domain structure to be followed visually as the crystal is taken through all stages of magnetisation parallel to the applied field. In a separate piece of apparatus the hysteresis cycle of a piece of steel is displayed by an oscilloscope. The question sheet prompts the linking of the hysteresis cycle with the motion of the domain walls and probes the connections between crystal quality, motion of the domain walls and the relationship between the different shape hysteresis curves and the applications of given materials.

Student guide

A material is ferromagnetic if internal magnetic interactions between constitutent atoms (possessing a magnetic dipole moment) align these moments.

A sample of ferromagnetic material, e.g. Fe metal, might be expected to be like

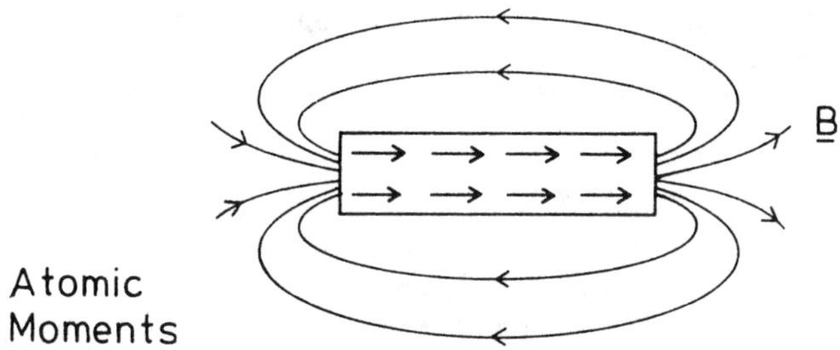

Generally this is not the case as such an arrangement produces a large magnetic field B outside the sample. To set up a field B costs energy $B^2/(2\mu_o)$ per unit volume.

A lower energy arrangement can be obtained if the atomic moments are aligned in sub-units (called domains) and the domains arrange themselves to minimise the energy.

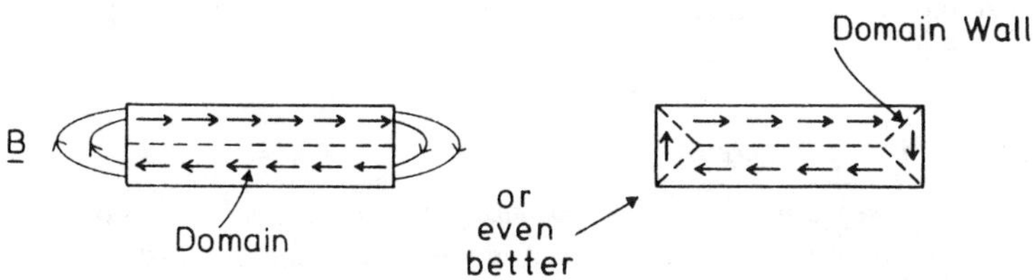

What arrangement of domains gives the lowest energy?

Total Energy =	$E_{\text{mag field}}$	$+E_{\text{anisotropy}}$	$+E_{\text{domain walls}}$	$+E_{\text{external field}}$
	↓	↓	↓	↓
	Sum of $\frac{1}{2}B^2/\mu_0$ over all space	within crystalline structure of material some directions of magnetisation have low energy (soft) parallel to field others high energy (hard)	costs energy to make domain wall	if external field present favours growth of domain with magnetisation

Observation of domains

Observe the domains in the single crystal sample of yttrium iron garnet ($Y_3Fe_5O_{12}$) as described on separate sheet (Magnetic domains). In this sample the magnetic domains are magnetised perpendicular to the plane of the crystal along the soft magnetic axis. Apply a magnetic field perpendicular to the plane of the sample.

Domains explain the hysteresis loop of ferromagnetic materials.

Hysteresis

An alternating magnetic field, H, is applied to a ferromagnetic sample by a current flowing through a coil. The total flux density B, which is composed of contributions from the coil and the magnetisation of the sample is measured by a second coil.

The relation between B and H is given by

B	=	$\mu_0(H$	+	$M)$	
↓		↓		↓	
total flux density		coil contribution		sample contribution	*Kip Fundamentals of Electricity and Magnetism,* p. 326.

A plot of B versus H gives the characteristic hysteresis loop of the sample.

Observation of hysteresis

Description of demonstration apparatus. The magnetic sample whose hysteresis is observed is the laminated iron yoke. The geometry is sketched in the diagram. The iron of the yoke is not a single crystal like the sample of YIG but is microcrystalline — composed of small crystals within the total block of iron.

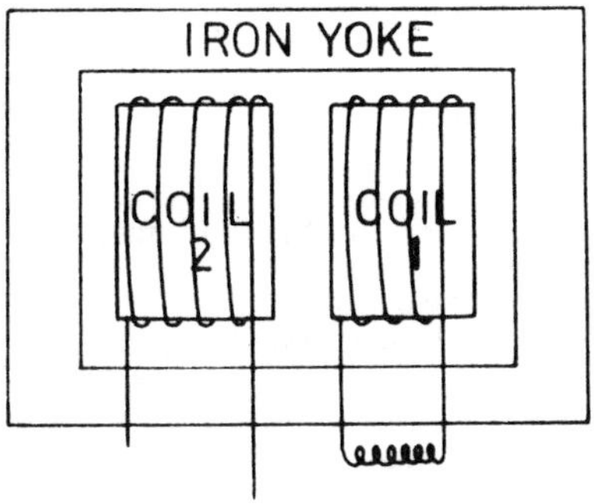

To observe the hysteresis connect signal B (green terminals onto channel 1 of the cathode ray oscilloscope and connect the signal H (blue terminals) to the external time base.

The hysteresis loop should look like

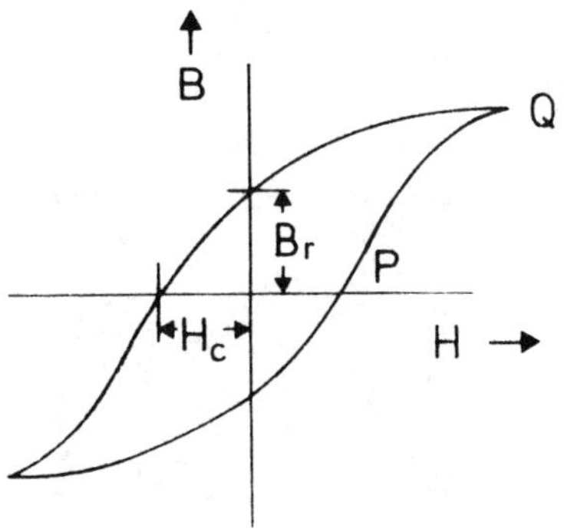

Bulk characteristics of such a loop are

(i) Remanence — value B_r remaining when H is reduced to zero from its maximum value.
(ii) Coercive force H_c — back value of H needed to reduce B to zero.
(iii) The area of the loop.

It can be shown (Kip: *Fundamentals of Electricity and Magnetism*, p. 364) that work done in one cycle on the magnetic sample is

$$\text{Work done} = \oint H.\mathrm{d}B = \text{area of loop}$$

The domain picture of the ferromagnetic sample provides an explanation of the shape of the hysteresis loop in terms of the adjustment of the magnetic domains to the external field. The motion of domain walls in a crystal is impeded and restricted by impurity atoms and dislocations of the lattice.

Try the questions on the sheet.

Further reading and reference (in Physics Library)

Boll: *Soft Magnetic Materials*.
Bobeck and Della Torre: *Magnetic Bubbles; Selected topics in Sold State Physics*.

Magnetic domains

In a very thin sample of ytterium iron garnet the magnetic domains can be made visible because they rotate the plane of polarisation of polarised light (Faraday effect) by different amounts.

The apparatus for this demonstration consists of a sample of this material positioned under the objective of a microscope between a pair of almost crossed polaroids. Light from a table lamp can be directed through the sample by adjusting the mirror on the bottom of the microscope. When focusing the microscope be careful not to squash the sample. The different domains should be clearly visible as shown in this diagram.

When you have seen the domains, confirm that they cannot be seen when the top piece of polaroid is removed or when the angle between the polaroid axes is significantly different from 90°. What happens to the domains when a magnet is brought near to the sample? A suitable magnet should be near the microscope.

Question Sheet

Q1 On YIG domain demonstration — explain briefly your observations and account for them in terms of the $E_{\text{mag field}}$ and $E_{\text{external field}}$ terms of the total energy relation.

Q2 Copy the sketch of the yoke and coils of the hysteresis demonstration — on this sketch the lines of B linking the two coils.

Hysteresis loop

Q3 From point P on loop B increases with increasing H but then almost saturates at Q. Give a brief explanation of this.

Q4 As H is decreased to zero from Q then B_r remains — give a brief explanation of this.

Q5 What features of a material makes for easy or difficult movement of domain walls?

Choice of suitable practical ferromagnetic materials

Q6 A transformer core acts to link the flux of primary and secondary coils. Draw the hysteresis loop of a material suitable for this.

Q7 Draw the hysteresis loop of a material that is suitable for magnetic recording (tapes) or computer memory.

Q8 Sketch hysteresis loop of a piece of machined iron — then sketch the changes if it is annealed (heated to release the internal strains). Which state of the iron would make the better permanent magnet?

Investigation 3 — Electron diffraction and the de Broglie wavelength

In this exercise electron diffraction patterns of powder and single crystal samples are observed and the scale of the patterns related to the electron wavelength. This wavelength is seen to be given by the de Broglie relation.

The apparatus is a large commercial cathode ray oscilloscope (Welch Corporation, Chicago) incorporating a wire mesh placed between the electron gun and the fluorescent screen. This mesh bears targets of polycrystalline aluminium and a single crystal of graphite with its six-fold symmetry axis oriented parallel to the electron beam. The fluorescent screen is flat allowing the patterns to be measured. Different accelerating voltages give rise to different electron wavelengths which are deduced and compared with those predicted by the de Broglie relation. Using a known electron wavelength the lattice constant of graphite is found.

Combined student guide and question sheet

This apparatus demonstrates that electrons have wavelike properties and can produce diffraction patterns when reflected from crystalline materials.

The electrons can be accelerated through potentials of up to 10 kV and focused onto the target situated halfway between the electron gun and a fluorescent screen; there is a spare tube for you to inspect the detail of the construction. There are two types of target, polycrystalline aluminium and pyrolytic graphite.

Switch on the power and allow a few moments for the set to warm up. Set the accelerating voltage to about 8 kV and slowly turn up the intensity, keeping an eye on the current meter alongside the set: try not to let the current exceed 20 μA. There should be a bright spot on the screen which can be moved about by adjusting the horizontal and vertical controls on the front panel. By careful hunting around the screen you should be able to find the two separate diffraction patterns, as here:

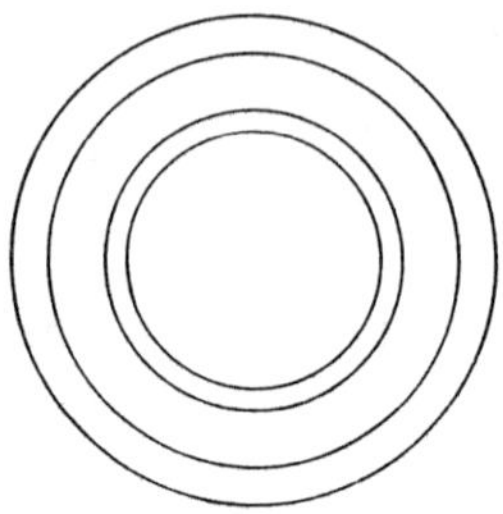

FIG. 1. Aluminium target.

FIG. 2. Carbon target.

It may help to fine tune the picture by small adjustments to the accelerating voltage V.

Analysis of Fig. 1

The polycrystalline aluminium target has a face-centre cubic lattice and the rings seen on the screen correspond to first order Bragg reflections from the planes with Miller indices 111, 200, 220, 311 (don't worry if you have not yet met Miller indices), starting with the innermost ring.

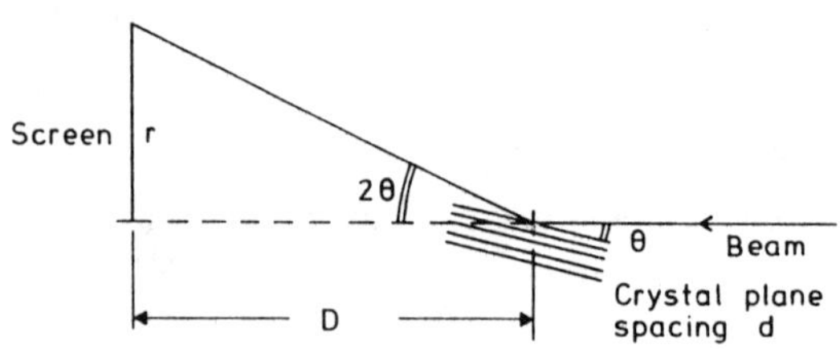

The Bragg condition is $2d \sin \theta = n\lambda$ where $2\theta = r/D$ radians. For this tube the screen-target distance $D = 18.29$ cm. (You measure r on the screen.)

For a cubic lattice the distance d between planes with Miller indices (h, k, l) is $d/n = a/(h^2 + k^2 + l^2)^{\frac{1}{2}}$, where a is the lattice constant which for aluminium is $a = 0.404$ nm (from X-ray data).

Measurement of powder diffraction rings

(i) Set the accelerating voltage to 8 kV and measure the radii of the four rings.
 a) Deduce a value of the electron wavelength λ from each ring.
 b) Calculate a value for the electron wavelength using the de Broglie relation and compare the values.

(ii) Using measurements of the innermost (111) ring, deduce values of the electron wavelength for accelerating voltages of 9 kV and 10 kV. In each case compare these values with those calculated using the de Broglie relation.

Analysis of Fig. 2

The pyrolytic graphite is made at high temperatures (hence the name) and the planes consist of hexagonal networks of aligned sheets of carbon atoms 10 to 100 atoms thick. The hexagonal symmetry of the structure is reflected in the diffraction pattern.

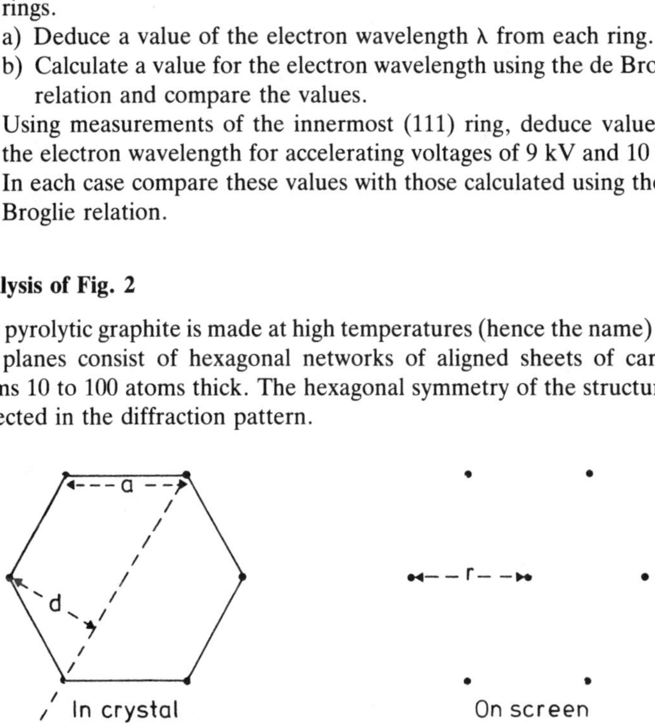

Measurement of single crystal spot pattern

Set the accelerating voltage at 8 kV and measure the distance r on the screen.

Using your value of the electron wavelength, the relation $2d \sin \theta = \lambda$ and the geometry of the diagram deduce a value of the lattice constant a for graphite.

Electron mass, m	$= 9.11\times10^{-31}$ kg
Electron charge, e	$= 1.6\times10^{-19}$ C
Planck's constant, h	$= 6.63\times10^{-34}$ Js

REFERENCES

1 A phenomena laboratory for physics students, *Phys. Educ.* **18** (1983), 125–127.

Fourier analysis of vowels

A.T. WINTER
Cavendish Laboratory (Cambridge University Physics Department), Madingley Road, Cambridge CB3 0HE

Abstract: The student is invited to sing a vowel into a microphone. The (periodic) waveform is displayed on an oscilloscope and is also digitised, stored in a microcomputer and displayed on its monitor. A BASIC program is then used to form the Fourier Series for the waveform — that is to calculate the amplitudes and phases of the fundamental and harmonics. The aim of this is to teach the student about Fourier series, and to make the connection between the formalism of Fourier notation and A-level teaching on the waveform of sound, its relation to the harmonics, and to the perceived timbre.

A separate program may then be used to reconstitute the waveform using the amplitudes and phases of the harmonics as input. Students may also study the production of other waveforms such as square and triangular waves.

The basic idea of analysing a periodic signal into a superposition of sinusoids (the fundamental and harmonics) is familiar to many students from work at school. University courses on applied mathematics and/or physics give a detailed treatment of the same idea using the formalism of Fourier analysis. It is a sad experience that many students do not realise that the two approaches are identical and they learn Fourier analysis in the abstract, possibly finding the concepts hard to grasp. The present experiment in the first year physics practical class attempts to bridge this gap, and also to guide the student through calculation of a particular Fourier series on the principle that the best way to learn a technique is to use it. The experiment involves digitising a waveform and storing it in a microcomputer: we have made an explicit point of observing the details of this process, but the instructions could be recast to make it more transparent.

Figure 1 shows a block diagram of the equipment used. The pre-

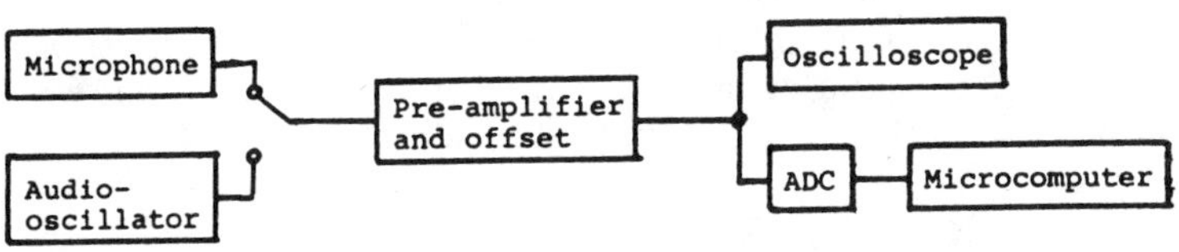

FIG. 1. Fourier series for the vowel 'a' as in bath, produced by a student

amplifier and analogue-to-digital converter were built specially. The microcomputer used is an Acorn Atom or a BBC model B.

The waveform is generated by singing a steady vowel into the microphone, or shy students may use a signal generator to produce an approximation to a square wave. The data is sampled with 8 bit precision at a sampling rate of 5 kHz. This is too fast for the analogue inputs built into the BBC Microcomputer but some fast commercial analogue-to-digital converters would be suitable. The author would be happy to supply full details of the equipment and instructions for carrying out the experiment.

The Fourier series is calculated using a short program written in BASIC to calculate the Fourier integrals directly. The option of using more powerful mathematical techniques was considered; for example far more data can be handled if the auto correlation function is calculated before the Fourier series is formed; alternatively a fast Fourier transform could be used. However these options were rejected in the interests of simplicity so that an enterprising student who printed out the BASIC programs would be able to understand them easily. Because the programs are not sophisticated they are slow, and not much data can be handled — perhaps three periods of a typical waveform (128 samples).

After the Fourier series has been calculated, a second BASIC program may be used to reconstitute the waveform. Alternatively other Fourier series may be summed converging towards waveforms such as a square, sawtooth or triangle wave. These programs are again simple and therefore understandable by students who think of looking at them and this is again regarded as an important point. The Fourier coefficients would be calculated by the student.

Frequency relative to fundamental	1	2	3	4	5	6	7	8
Cosine amplitude	.243	−.073	−.082	−.027	−.057	−.055	−.015	−.010
Sine amplitude	.000	−.181	−.115	.030	.079	.068	.063	.073

A collection of over a hundred waveforms produced by different students supposedly singing the same vowel showed wide variations which might repay analysis on a geographical basis!

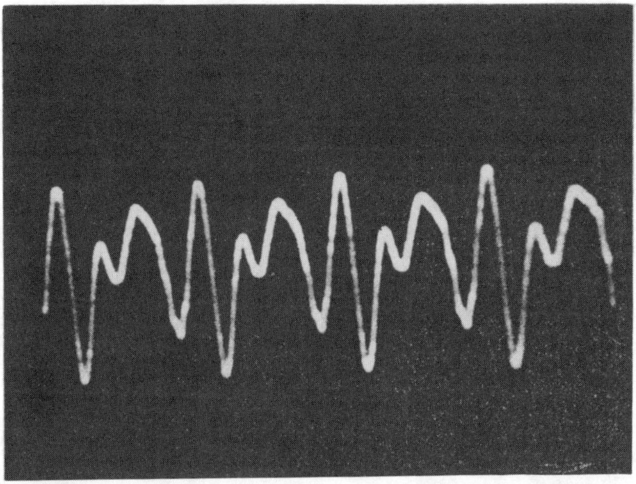

FIG. 2. Waveform for the vowel 'o' in rose, sung by the author and photographed from the screen of a storage oscilloscope.

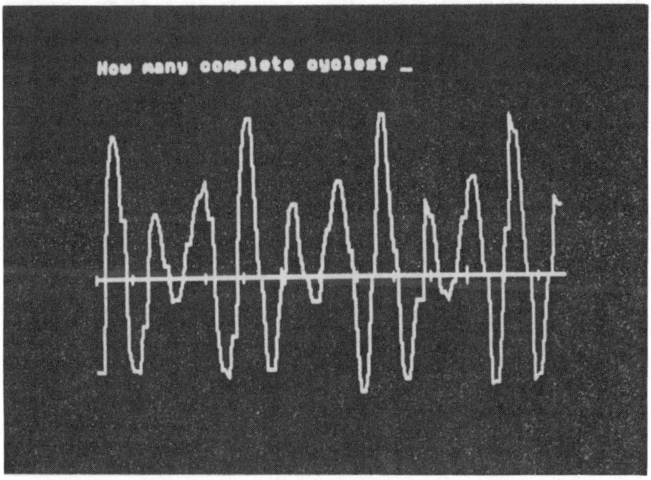

FIG. 3. The same vowel as Fig. 2 but digitised and photographed from the monitor of a BBC Microcomputer. Differences between Figs. 2 and 3 are genuine and indicate the difficulty of producing the same waveform on two separate occasions. The computer's question 'How many complete cycles?' refers to the first stage of calculating the Fourier series.

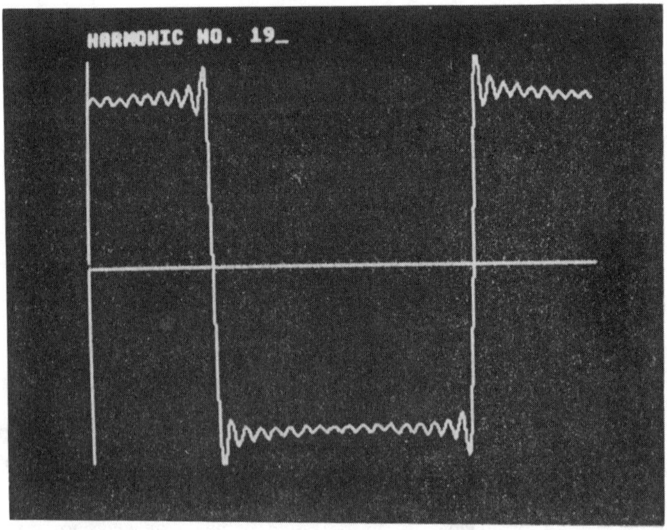

FIG. 4. Fourier series for a square wave truncated after nineteen harmonics. The Gibbs phenomenon is well developed. The series consists of odd harmonic cosine terms with amplitudes inversely proportional to frequency and alternating signs — i.e.

$$\cos(t) - (\cos 3t)/3 + (\cos 5\,t)/5 - (\cos 7t)/7 + \text{etc.}$$

The figure is obtained by photographing the monitor of the BBC Microcomputer.

Reflection of ultrasound by boundaries between liquids

P.H. Pope and A.T. Winter
Cavendish Laboratory (Cambridge University Physics Department), Madingley Road, Cambridge CB3 0HE

Abstract: Pulses of sound are propagated through liquids and reflected from liquid–air and liquid–liquid interfaces. The speed at which the pulses travel is measured by the time-of-flight method and the densities of the liquids by use of a hydrometer. The amplitude reflection coefficient for the interface can then be calculated and is found to agree well with the measured value. An alternative development would be to study reflection of the sound from small solid objects suspended in the beam.

Propagation of sound through a fluid is controlled by the density and the bulk modulus. For present purposes we need to define the wave velocity c and the fluid density ρ. the product of these is knows as the acoustic impedance $Z = c.310$ (unit kg m^{-2} s^{-1}).

For liquids the density may be measured quickly and accurately by using a suitable hydrometer. In the present experiment the wave velocity is found by measuring the time of flight of pulses of sound at a frequency of 670 kHz. The pulses are launched upwards by a piezoelectric transducer placed at the bottom of a glass cylinder containing the liquid. The cylinder has the dimensions, roughly, of a 500 ml measuring cylinder. Each pulse travels up through the liquid, is reflected by the liquid–air interface, and is detected when it arrives back at the transducer. The time of flight may be measured on an oscilloscope or, preferably, on a digital timer/counter. The pulse is partially reflected by the transducer and may therefore be detected many times as it bounces up and down the column of liquid. Strictly speaking this method gives the group velocity rather than the wave velocity but since simple liquids are not dispersive at the frequency used this distinction is not important.

The acoustic impedance is found simply by multiplying together the measured values for wave velocity and density.

An interface between two media causes partial reflection of an incident sound wave if the acoustic impedance changes. The amplitude reflection coefficient R is given by the formula

$$R = (Z_2 - Z_1)/(Z_2 + Z_1)$$

Reflections of this type arise from changes in the density of sea water and can cause confusion when ultrasonic beams are used for depth-finding or

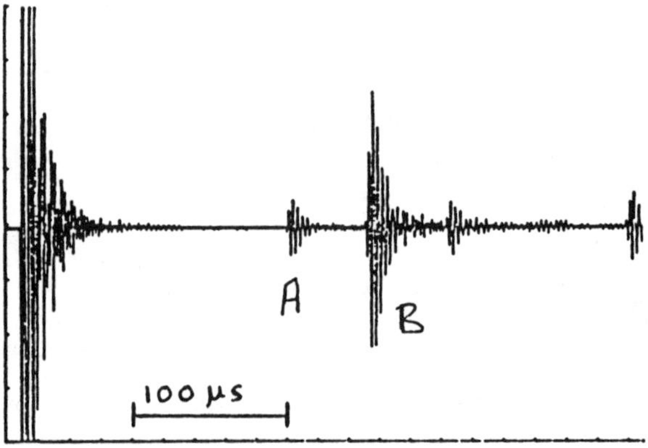

FIG. 1. Output from the transducer captured using a transient recorder. The trace shows the original departure of a pulse and its arival back at the transducer after reflection by a water–paraffin interface (A) and a paraffin–air interface (B).

for detection of objects underwater. The phenomenon is illustrated by floating a thick layer of paraffin onto water in the apparatus. This results in a weak pulse from the water–paraffin interface (R = 0.17) preceding the stronger pulse from the paraffin–air interface. By comparing the amplitude of the weak pulse with that of the reflection from the water–air interface (measured before adding paraffin to the apparatus) an experimental value for R may be obtained and compared with the expected value calculated above.

If a small solid object is suspended in the liquid below the surface then the sound reflected from it may be observed. This can lead to a discussion of ultrasonic non-destructive testing of engineering and metallurgical specimens; sound is reflected strongly by sub-surface flaws such as cracks and inclusions.

Apparatus required for this experiment consists of an oscilloscope, an audio oscillator giving a square-wave of amplitude about 0.5 V and frequency about 300 Hz, and if possible a digital counter/timer. Additionally a transducer is required to launch and detect the pulses and a special circuit is required to drive the transducer. This additional equipment costs perhaps £50 and full details are given by Pope and Winter[1] or ATW would be able to provide the specification to anyone who is interested.

REFERENCE

1 P. Pope and A.T. Winter: *Eur. J. Phys.* **6,** 13 (1985).

HEAT AND THERMODYNAMICS

Pressure-volume relations for freon and air

W. Vennart and J.R. Drabble
Physics Department, University of Exeter, Devon

Abstract: An apparatus for the measurement of pressure–volume relations for freon (dichlorodifluoromethane) and air is described. For air, pressure is found to be inversely proportional to volume over the pressure range (1×10^5 to 6.5×10^5) Pa. Freon follows a similar relationship up to a pressure of approximately 5.5×10^5 Pa where it liquefies, and further reductions in volume result in a constant pressure measurement equal to the saturated vapour pressure of freon at the particular temperature. A series of isotherms for freon have been measured up to a pressure of 7.0×10^5 Pa and a temperature of 310 K.

There have been many reconstructions of Andrews'[1] experiment involving the measurement of pressure–volume relations for carbon dioxide which in general necessitate the use of high pressure apparatus with suitable safety shielding. In this experiment the ideal gas equation:

$$PV/T = \text{constant}$$

can be investigated for both freon and air up to pressures of 8.0×10^5 Pa. In particular, freon, which has a critical temperature of 385 K, exhibits significant departures from this relationship. Air exhibits approximately ideal behaviour up to this pressure. In addition variations of the pressure–volume (P, V) isotherms with temperature can also be investigated (Fig. 1).

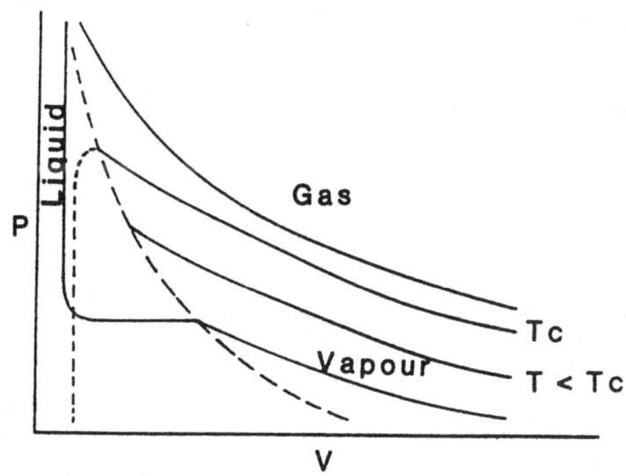

Fig. 1. Pressure (P) plotted against volume (V) for a particular substance at different temperatures — T_C is the critical temperature.

The experimental chamber is a vertical uniform glass tube (internal diameter 4.5 mm, external diameter 6 mm and length 200 mm) in which freon and air can be separately enclosed. The bottom of the tube is attached to a mercury reservoir. The pressure and volume of the gas enclosed in the tube can be varied by admitting mercury into the tube with the aid of a screw-driven pump. Figure 2 shows a schematic of the apparatus, the majority of which is made of stainless steel. The graduated glass tube, for enclosing the gases, is attached at its lower end to a reservoir of mercury (which is terminated by metal bellows) and at its upper end to a needle valve by 'O'-ring seals. The needle valve enables either freon (contained in a small cylinder) or air to be admitted into the glass tube. The height of the mercury in the tube can be varied by altering the length of the metal bellows contained in the screw-driven pump — in this way the pressure and volume of the gas in the tube can be changed. The pressure of the gas in the tube is measured using a 'Budenberg' gauge and its volume from the graduations on the side of the glass tube (which is assumed to have uniform cross-sectional area). The pressure measured by the gauge has to be corrected for the height of the column of mercury in the glass tube and any zero errors. The temperature of the gas in the tube can be varied by circulating water, at a constant temperature, through the perspex tube surrounding the system.

Typical curves of pressure (*P*) versus volume (*V*) and inverse volume (1/*V*) at room temperature (297 K) are shown in Figs 3 and 4 for air and freon. Over the range of pressures used air shows a linear relationship between *P* and 1/*V* (the correlation coefficient for the line is 0.998), thus

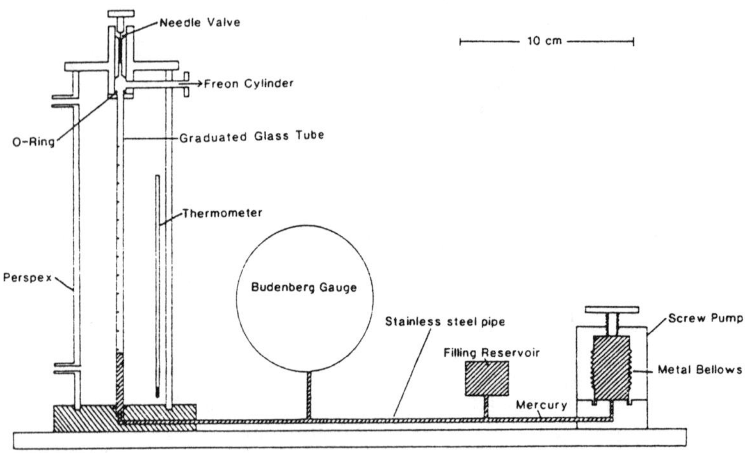

FIG. 2. Schematic diagram of the apparatus.

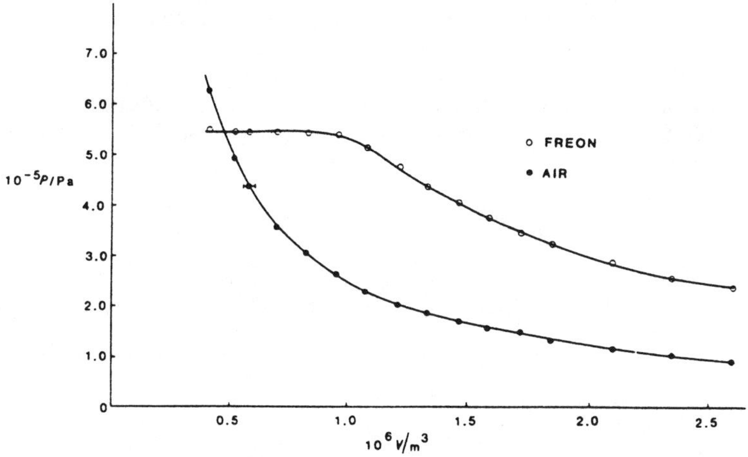

FIG. 3. Pressure (P) versus volume (V) for air and freon at room temperature (297 K). Freon reaches a saturated vapour pressure of approximately 5.5×10^5 Pa.

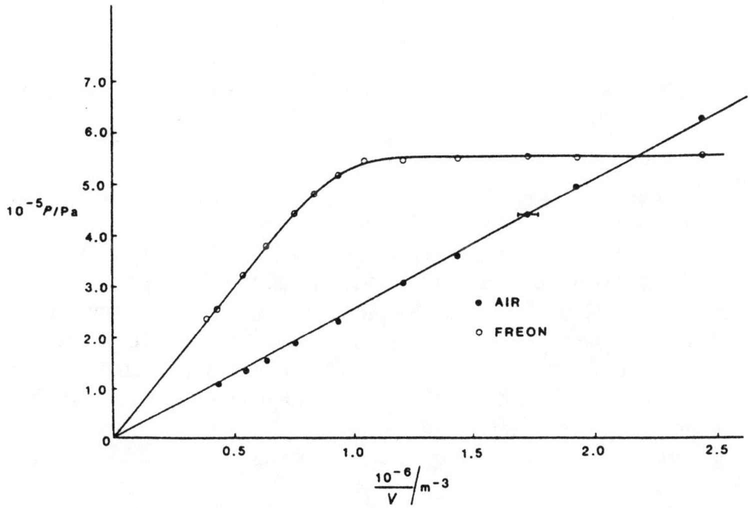

FIG. 4. Pressure (P) versus reciprocal volume ($1/V$) for air and freon at room temperature (297 K). Freon obeys the ideal gas equation up to a pressure of approximately 4×10^5 Pa.

verifying the above equation. Freon only obeys the equation up to a pressure of approximately 4.0×10^5 Pa — at approximately 5.5×10^5 Pa the pressure measured is that of the saturated vapour. A typical set of isotherms for freon are shown in Fig. 5 — further modification to the apparatus will have to be made if pressures greater than 8×10^5 Pa are to be used.

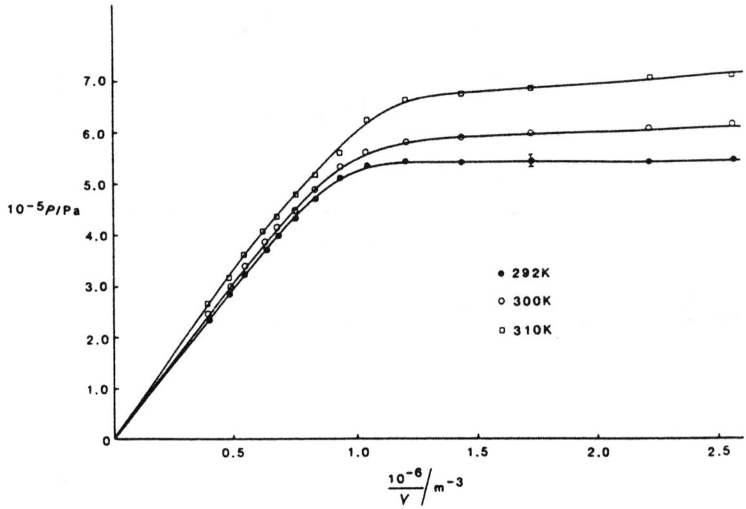

FIG. 5. Pressure (P) versus reciprocal volume ($1/V$) for freon at temperatures of 292 K, 300 K and 310 K.

The apparatus described above was originally reported by Karas and Davis[2]. Some major modifications have been made to the system described here to obtain reproducibility of results. It is relatively easy and cheap to construct using normal workshop facilities and the freon cannisters can be obtained from most suppliers of refrigeration equipment. In our experience undergraduates find the experiment interesting because it demonstrates some of the basic physics of gases which they rarely have the opportunity to experiment on. The data obtained is reproducible, e.g. students report that freon liquefies at the same pressure, for a given temperature, to within a range of approximately 5%.

REFERENCES

1 Andrews: *Phil. Trans.* **575** (1869) and **421,** 167 (1867).
2 R.H. Karas and P.D. Sumner: *American Journal of Physics* **47** (5), 473 (1979).

A simulation of an order–disorder phase transition using a microcomputer

C.M. CARE
Department of Applied Physics, Sheffield City Polytechnic, Pond Street, Sheffield S1 1WB

Abstract: The standard Metropolis Monte-Carlo method is used to simulate the temperature dependence of the degree of ordering in a three-state Potts model. This mathematical model can represent a variety of real physical systems such as liquid crystals or ferromagnets and exhibits a first order phase transition.

The simulation employs a simple cubic lattice of 8×8×8 with periodic boundary conditions. The program runs on the Dragon 32 microcomputer and is written principally in 6809 assembly language in order to achieve the necessary speed.

The computer 'experiment' displays the state of ordering on the graphics display and also outputs data which can be used to determine the order parameter at each temperature. The students can investigate the order parameter as a function of temperature. The number of Monte-Carlo steps can be varied and this is particularly important near the phase transition. The results obtained are close to those obtained using much larger lattice arrays.

Introduction

Many physical systems exhibit phase transitions between a high temperature disordered phase and a low temperature ordered phase. This ordering may be in the position of the particles as, for example, in a liquid to solid transition. Alternatively, the ordering may occur in the orientation of the particles in the system. This type of ordering is observed in the paramagnetic to ferromagnetic phase transition. The molecules in liquid crystal phases can exhibit both types of ordering.

Three state Potts model

In order to study the properties of order–disorder phase transitions, it is useful to develop mathematical models which exhibit such phase transitions. One such model is the three state Potts model.[1,2] In this model, we assume that the points of a lattice are each occupied by a 'spin' which can only align itself in one of three mutually perpendicular directions (see Fig. 1). These 'spin' directions may represent the magnetic orientation in a magnetic system or the orientation of a molecule in a liquid crystal. It is assumed that each spin will interact only with its nearest neighbours and that there will be an energy $-J$ associated with every pair of spins which are parallel. Thus the minimum energy of the system will be a state in

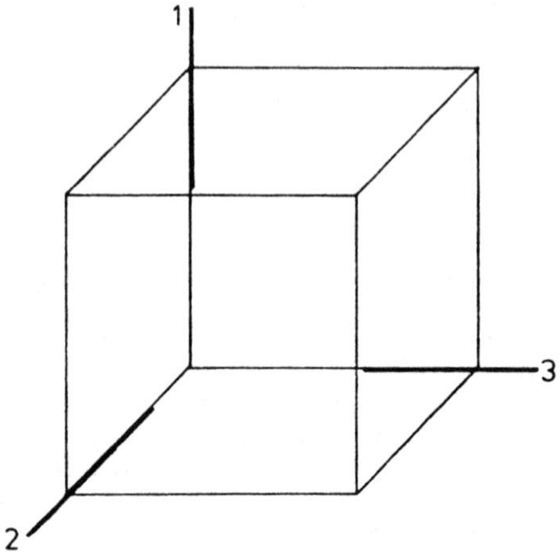

FIG. 1. Possible spin orientations in three state Potts model.

which all the spins are aligned in one direction. The energy J is known in magnetic systems as the *exchange interaction*. We may write down the total energy of the system, H, as

$$H = -J \sum_{<ij>} \delta_{S_i S_j} \tag{1}$$

where the sum is over all nearest neighbour pairs in the lattice and $S_i = 1$, 2 or 3, where S_i is spin direction on site i

$$\delta_{S_i S_j} = 0 \qquad S_i = S_j$$

$$\delta_{S_i S_j} = 1 \qquad S_i = S_j$$

Equation (1) represents the Potts model. We may use the formula to calculate the energy of any arrangement of the spins. If a particular arrangement has energy E_k, the probability of it occurring at a temperature T is proportional to the Boltzmann term

$$\exp \{-E_k/kT\} .$$

If we consider all possible states of the system, we can in principle calculate the most likely energy and degree of ordering of the system at any temperature. Unfortunately, even this simple model cannot be solved analytically for a three-dimensional lattice and we must use various approximations to study the model. An alternative technique is to use a computer to simulate the behaviour of the system using a 'Monte Carlo' method.

Monte Carlo technique

We use the Metropolis[3] Monte Carlo technique, which is a standard method for counting configurations of a system at a given temperature. In this 'experiment' we consider an 8×8×8 array of spins on a simple cubic lattice with periodic boundary conditions. Thus the array of spins is assumed to be repeated indefinitely in all directions.

In one Monte Carlo *MOVE* the following procedures are followed:–
(i) a spin is chosen at random from the 512 spins in the array;
(ii) a new orientation for the spin is chosen at random;
(iii) the change in energy of the system, ΔE, is calculated assuming the spin changes from the old to the new orientation;
(iv) If ΔE is negative, the new orientation is accepted;
(v) if ΔE is positive, the new orientation is accepted with probability $\exp\{-\Delta E/kT\}$.

A Monte Carlo *STEP* (MCS) is made up of a large number of these MOVES. Thus 512 MOVES would allow an average of one move per spin. In this experiment the number of moves per step can be varied but *it is suggested that a value of 512 moves per step* is used throughout. After each STEP the number of spins in each of the three directions is counted. A large number of MCS are made at each temperature to determine the average number of spins in each direction at that temperature. Before collecting data at any temperature a number of MCS must be run to allow the array to reach thermal equilibrium at that temperature. Thus, if 400 MCS are taken, the first 100 may be ignored as 'thermalisation' steps. At each temperature we may calculate an *ORDER PARAMETER*

$$\xi = \tfrac{1}{2}\{[3N_{max}/(N_1+N_2+N_3)] - 1\}$$

where N_1, N_2, N_3 are the total number of spins in each direction and N_{max} is the largest value of N_1, N_2 and N_3.

The order parameter will be zero for the disordered state ($N_1 = N_2 = N_3$) and one for a fully ordered state (e.g. $N_1 = N_2 = 0 \quad N_3 = N_{max}$).

Experiment

The programme on the Dragon 32 allows the investigation of the behaviour of a three state Potts model using this Monte Carlo method. It is written principally in assembly language to achieve the necessary speed. The temperature is quoted in the dimensionless units of kT/J. The programme allows the student to determine the order parameter as a function of temperature and typical results are shown in Fig. 2. Six of the lattice planes are also displayed on the graphics screen and the order–disorder transition can be clearly observed.

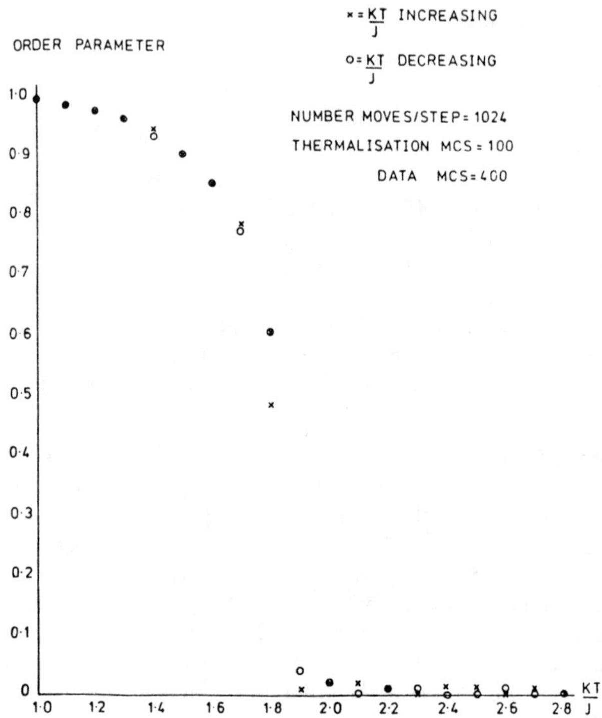

FIG. 2. Typical results showing variation of order parameter as a function of kT/J.

Further details of the programme can be obtained from the author at the above address.

REFERENCES

1 R.B. Potts: *Proc. Camb. Phil. Soc.* **48**, 106–109 (1952).
2 M.W. Blote and R.H. Swensen: *J. Appl. Phys.* **50**, 7382–7384 (1979).
3 N. Metropolis, A.W. Rosenbluth, N.M. Rosenbluth, A.H. Teller and E. Teller: *J. Chem. Phys.* **21**, 1087–1092 (1953).

A three point temperature display and data logger

J. WOODMAN
Sheffield City Polytechnic, Department of Biological Sciences, Pond St., Sheffield S1 1WB

Abstract: The design of an economical and practical platinum resistance thermometer for the BBC Microcomputer is described briefly. Its accuracy is better than ±0.5°C and it is suitable for use in schools as a display thermometer for class demonstrations and as a data logger.

This is a project which could be managed by most sixth form physics students. It provides the school with three platinum resistance thermometers for use in demonstrations and practical classes. The recorder can easily be made to provide a graphical record on a TV screen and/or a file of results on disc or tape.

Even though the full cost of converting your BBC Model B to such a three point temperature recorder is low it is extremely flexible in use. By connecting the computer to a large TV it provides a continuous record of the temperature clearly visible to all the pupils in the class. If a file of time and temperature readings is made during an experiment then a separate program can be prepared which will give an immediate graph of the results as well as an error proof calculation! The quick turn round of results is useful in allowing an experiment to be repeated immediately if necessary.

Staff in subjects other than physics and chemistry such as home economics or environmental studies could well find the equipment useful, not only as a temperature recorder but also to demonstrate the technique of data logging.

Principle and expected accuracy

The analogue input port measures voltages in the range 0 to 1.8 volts. The circuit in Fig. 1 shows a resistance thermometer of 1000 ohms at 0°C in series with a standard resistance of 3k ohms. The potential at point A at

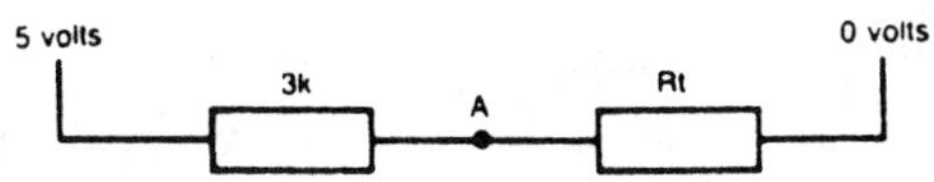

FIG. 1.

0°C would be 1.25 V. At 100°C the resistance of the thermometer will have increased to nearly 1400 ohms and the potential to almost 1.6 V.

The analogue to digital converter gives the computer a number which is proportional to the potential at the port. Generally an input of 1 V produces a reading of 35168. This is a 16-bit number (range 0 to 65520) but the converter only gives 12 bits. Of these 12 bits only the first 10 are reliable: effectively the A-D converter is working on a scale of 0 to 4096 with an accuracy of 1 in 1024 although the numbers used are in the range 0 to 65520.

Thus the resistance is theoretically accurate to better than 2 in 100. The temperature would be subject to an error of between 0.5 and 1°C depending on the temperature being measured. These are theoretical calculations and only represent the best that can be expected under ideal conditions.

In practice the readings obtained wil this circuit drift uncontrollably for several hours even after the computer has warmed up. The circuit in Fig. 2 uses the potential at point S as a standard to determine the potential at point A. This provides a measure of potential which remains stable over several hours after allowing the first few minutes for the computer to stabilise.

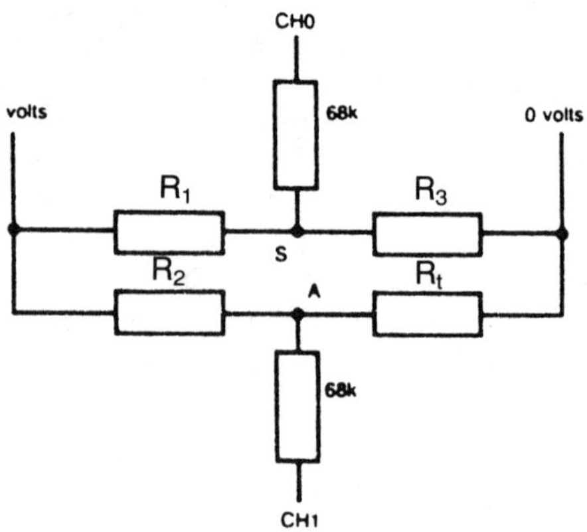

FIG. 2.

In this circuit

$$\text{reading } (S) = k.\text{volts}.R_3/(R_1+R_3)$$

$$\text{reading } (A) = k.\text{volts}.R_t/(R_t+R_2)$$

$$\text{ratio} = \text{reading } (S)/\text{reading } (A) = R_3.(R_2+R_t)/((R_1+R_3).R_t)$$

$$R_t = R_2/((\text{ratio}/S)-1)$$

where $S = R_3/(R_1+R_3)$

Note that this measure of the resistance is independent of the actual value of the voltage used. Values for the resistance determined in this way are steady and reproducible. They are, however, very dependent on the absolute values of the standard resistors in the circuit. In the model which we use in our food processing pilot plant we have selected 0.4W metal film resistances because of their low temperatures coefficient of resistance and because they are available as 1% grade. Wire wound high precision resistors are available which are accurate to 0.1%. As these cost several pounds each their cost cannot be justified with this circuit.

The relationship between the resistance of platinum and temperature is well known and given in all 'A' level physics books.

$$t/100 = (R_t - R_o)/(R_{100} - R_o)$$

where t = temperature, R_t etc. the resistance at t, zero and 100°C.

This relationship is so reliable that it has been used as a secondary thermometric standard for as long as there have been thermometric standards. The use of the computer allows the observed potential to be converted into temperature using exact arithmetical relationships. The fact that the relationship is non-linear presents no problem unlike the usual digital readout instrument where the choice of transducer has, in the past, been limited to those giving a linear response. The computer avoids the need to make do with second best!

In use it is essential to decouple the external circuitry from the computer; a resistance between 50k and 100k ohms does this quite satisfactorily. The resistor only increases the apparent internal resistance of the analogue port which our theory assumes to be infinite.

Listing 1 will convert the input at the analogue input port into temperature. It permits the exact values of the resistances used to be applied in the calculation. Note that ADVAL(0) is used to find out when conversion of channel 1 is complete. Firstly ADVAL(1) is read to clear the register for ADVAL(0). ADVAL(0) will then return a value of 256 next time this channel completes conversion (see *User Guide*, p. 203). The channels are converted in reverse numerical order so that completion of channel 1 indicates that all channels have been converted. The program immediately reads all four channels. The amendments in listing 2 give a graphical output so that the variability is easily seen. Note that listing 2 is not a program but must be typed in to amend listing 1.

The accuracy achieved in this version of the program is not good enough for general use. Where a variable is randomly changing around its mean, the standard error of the measure can be greatly improved by taking the mean of several readings.

$$SE = SD/\sqrt{N}$$

where $\sqrt{N}$ = the square root of the number of readings.

Thus the taking of the mean of four readings reduces the standard error to half of the standard deviation and the mean of nine readings gives a third of the standard deviation. The program can be easily modified with a FOR-NEXT loop to average different numbers of readings and to observe the effect on the reproducibility of the temperatures. It is also possible to

```
100REM LISTING-1
110@%=&20105
120DIMreading(4)
130DIMtemperature(4)
140Rzero=1000:Rhund=1386
150R=100/(Rhund-Rzero)
160R1=3000
170R2=3000
180R3=1000
190S=(R1+R3)/R3
200CLS
210PROCread
220TIME=0
230REPEAT
240UNTIL TIME>500
250GOTO 210
260 ********************************
270DEFPROCread
280W=ADVAL(1)
290REPEAT
300UNTIL ADVAL(0)=256
310FOR Q=1 TO 4
320reading(Q)=ADVAL(Q)
330NEXT
340FOR Q=2 TO 4
350reading(Q)=reading(1)/reading(Q)
360NEXT
370FOR Q=2 TO 4
380res=R2/((reading(Q)*S)-1)
390temperature(Q)=(res-Rzero)*R
400NEXT
410 FOR Q=2 TO 4
420 PRINTTAB(10,10*(Q-2))CHR$(141)temp
erature(Q)
430 PRINTTAB(10,(10*(Q-2))+1)CHR$(141)
temperature(Q)
440NEXT
450ENDPROC
```

LISTING 1.

```
100REM LISTING-2
200INPUT"Which channel do you want to
display ?   ";Ch
205MODE 0
240UNTIL TIME>5
250IFX<1280 GOTO 210
257END
410X=X+2
420DRAWX,temperature(Ch)*10
430REM 430 must be deleted
440REM 440 must be deleted
```

LISTING 2.

experiment with taking the means of the ADVAL readings rather than of the tempertures.

The actual results of this type of experiment depend on the quality of your soldering and the accuracy of the values for the resistors used. Our software takes the mean of 20 readings making our prototypes generally reproducible to 0.3 degree and accurate to better than 1 degree throughout the range. Note that the resistance of contacts and soldered joints are known to vary with time so that regular checks of accuracy are needed.

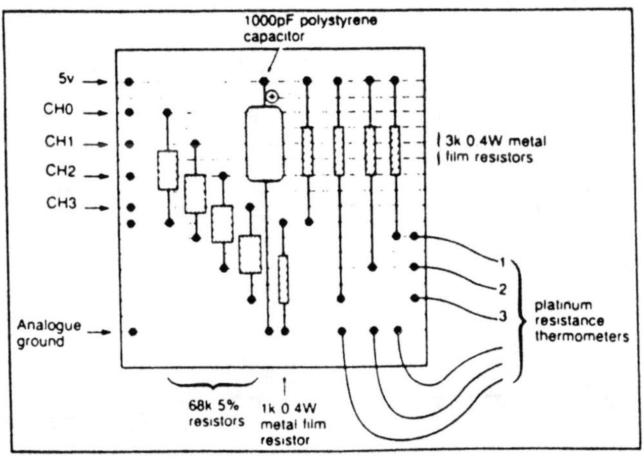

FIG. 3. The connections on a piece of Veroboard.

Preservation by heat processing

J. WOODMAN

Sheffield City Polytechnic, Department of Biological Sciences, Pond St., Sheffield S1 1WB

Abstract: A practical application of the platinum resistance thermometer operated by the BBC Microcomputer is described. It involves integrating temperature by time in real time. This paper is used as a class handout on a degree course for management in the Food Processing Industry.

Investigations into bacterial death in moist heat from 1910 onwards all demonstrated that the number of viable cells reduced exponentially with time at constant temperature, any deviations being slight and limited to the first few minutes. Although the exact physical interpretation of this relationship is still being discussed the so-called logarithmic relationship itself is not in doubt. The proportion of survivors in a medium at constant temperature can therefore be fully described by a single constant derived from the standard equation for first order kinetics.

The meaning of 'exponential relationship' is that a plot of the logarithm of the number of survivors against time gives a straight line as shown in Fig. 1. The number of survivors in that figure are 1000, 100 and 10 at 5, 10 and 15 minutes respectively. The time taken to reduce the number of viable cells present to one tenth of its original number is a constant (in this case 1 minute) independent of the number of cells present. It is called the decimal reduction time (D).

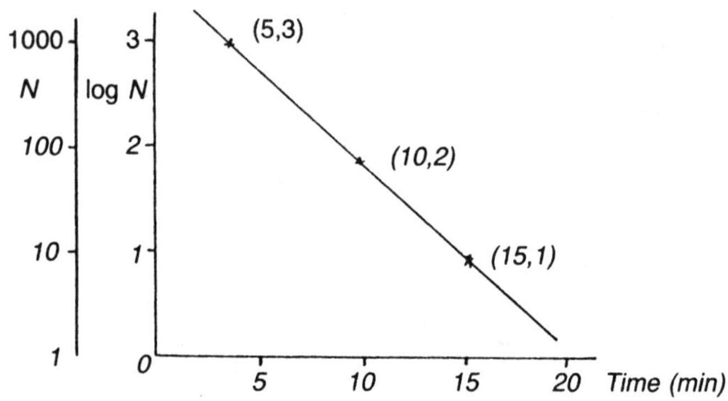

FIG. 1.

If the number of viable cells is N, the rate of loss may be written as the differential $-dN/dt$. Its proportionality to N then gives the equation:

$$-dN/dt = kN \quad \ldots \text{ k is a constant and } t \text{ is time.}$$

$$-dN/N = k dt$$

$$-\int dN/N = k \int dt$$

$$-\ln N = k\, t$$

$$-\ln N_2 - (-\ln N_1) = k(t_2 - t_1)$$

$$k = (\ln N_1 - \ln N_2)/(t_2 - t_1)$$

converting to logs base 10

$$k = 2.303/(t_2 - t_1) * \log (N_1/N_2)$$

The decimal reduction time is $(t_2 - t_1)$ when $N_1 = 10 * N_2$; substituting gives the relation between k and D.

$$k = 2.303/D * \log 10 = 2.303/D$$

The decimal reduction time depends on (1) the bacterium, (2) the strain of that bacterium, (3) the medium in which the cells are heated, and possibly (4) the prior treatment of the culture. The great bulk of this work was carried out on the effect of moist heat on spores, the details cannot be extended beyond that limitation. Although the original work was limited to a maximum of 130°C the results are frequently extrapolated beyond the limit; decisions based on such data must be interpreted with caution.

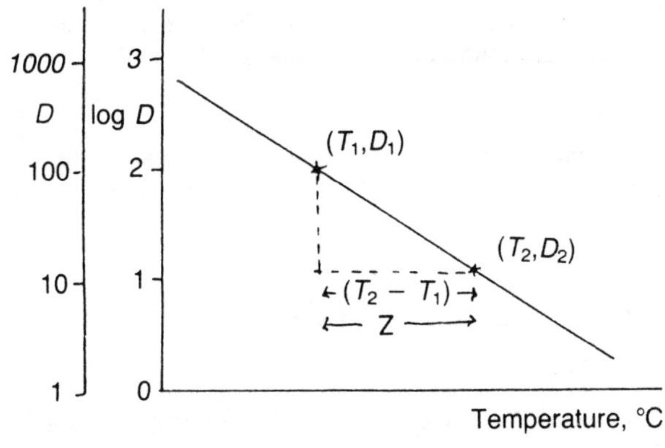

FIG. 2.

It is possible to determine the decimal reduction time at a range of temperatures. As the temperature increases so the decimal reduction time decreases, a graph of log (D) against temperature gives a straight line showing that this relationship too is exponential (see Fig. 2). This graph is called the thermal death time (TDT) graph. Straight lines have equations of the form:

$$y = m*X + C$$

$$\log(D) = m*T + C$$

substituting the two points (D_1,T_1) and (D_2,T_2)

$$\log(D_1) = m*T_1 + C$$

$$\log(D_2) = m*T_2 + C$$

subtracting

$$\log(D_1) - \log(D_2) = m*(T_1 - T_2)$$

$$\log(D_1/D_2) = m*(T_1 - T_2)$$

choosing $D_1 = 10*D_2$

$$\log(D_1/D_2) = 1 = m*(T_1 - T_2)$$

$$m = 1/(T_1 - T_2)$$

Note that $(T_1 - T_2)$ is negative as T_2 is the higher temperature. The positive value of this constant is called 'z', which together with a single value of D at a defined temperature, fully defines the relationship. As we chose $(D_1/D_2) = 10$, z becomes the rise in temperature needed to reduce the decimal reduction time by a factor of 10; it is usually taken as 10, the value determined in 1922, but recent work suggests that a figure of 7 or 8 may be closer for C. botulinum types A and B. Putting $-z = (T_1 - T_2)$

$$\log(D_1/D_2) = (T_2 - T_1)/z$$

In order to compare the effects of heating at different temperatures we can consider the thermal death time curve slope z passing through $D = 1$ at 121.1°C.

$$T_1 = 121.1 \qquad D_1 = 1$$

$$\log(1/D) = (T_2 - 121.1)/z$$

$$(1/D) = 10^{(T_2 - 121.1)/z}$$

Thermal processes for foods are designated as being five times D or 12 times D ($5D$ or $12D$ processes) where the factor 5 or 12 is decided in each individual case to give the required safety margin. clearly the safety margin for a pathogenic organism must be larger than that which is acceptable for a spoilage organism. This will be expressed as a number of minutes at 121.1°C and called the *F* value. In real life processes the temperature and hence *D* value does not remain constant and the difficulty of this work is to calculate back to equivalent minutes at 121.1 when the temperature is constantly changing.

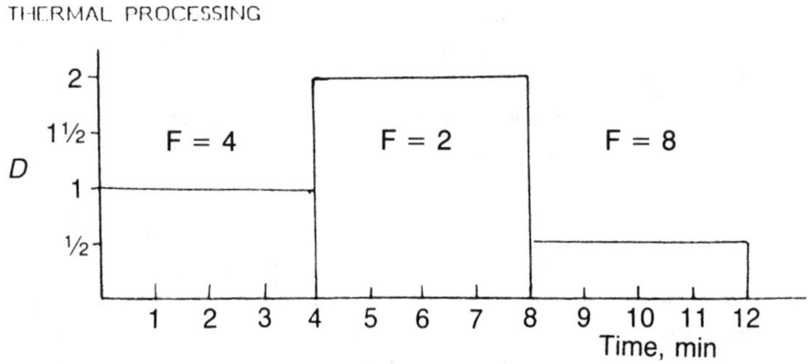

FIG. 3.

Figure 3 shows a hypothetical relationship between *D* value and time. In this case each of the first 4 minutes is worth 1 minute as the *D* value is 1; the next 4 minutes pass whilst the *D* value is 2 (the temperature has dropped) and each minute is therefore worth only 0.5 minutes; similarly each of the last 4 minutes must be counted as 2 minutes as the temperature has risen again to give a *D* value of 0.5. What is being done is to sum $t \times (1/D)$ arriving at a number of minutes at 121.1°C which would give the equivalent degree of sterility, in this case 14. Mathematically speaking this is integration and can be written:

$$F = \sum_{0}^{t} (1/D).\, \mathrm{d}t$$

$$F = \sum_{0}^{t} 10^{(T-121.1)/z}.\mathrm{d}t$$

If a record is made of temperature during a process then each temperature may be converted into an equivalent *D* value and the equivalent process at 121.1°C calculated. This may be done by dividing the temperature curve into equal time divisions and treating the mean temperature in each division as if it were constant. The smaller the divisions used the more accurate the process will be. Instruments are now available that do this integration during the process giving a continuous readout of *F*; indeed it is simple to write a program to make a home computer do this task accepting a temperature input from an analogue port.

The stages in the calculation are (1) determine the temperature, (2) find D at this temperature, (3) calculate the time at 121.1 which would give an equivalent process, and (4) sum these times.

The normal margin of safety for heat processing dangerous organisms is 12 times the D value — a $12D$ process: this means that a container having a single organism at the start of the process would have a 1 in 10^{-12} chance of a viable organism remaining at the end of the process. A 'proper' heat process is arrived at by deciding the organism that it is desired to eliminate and the safety factor that is acceptable. For C. botulinum this is a $12D$ process but for the flat sour organisms such as Bacillus stearothermophilus a $4D$ process may be considered acceptable. Table 1 gives some D values and F values for organisms important in heat processed foods.

TABLE. 1. F and D values for various organisms.

B. stearothermophilus	flat — sour	$D_{121} = 4–5$	$F_o = 8–12$
C. nigrificans	sulphide smell	$D_{121} = 2.5$	$F_o = 6–8$
C. botulinum	botulism	$D_{121} = 0.1–0.2$	$F_o = 2.5$
B. coagulans	spoilage	$D_{121} = 0.01$	
C. pasteurianum	butyric	$D_{100} = 0.1$	$F_{100} = 1$
Leuconostoc spp	fruit spoilage	$D_{66} = 0.7–1$	$F_{85} = 1$

Experiment

Prepare two lots of 3 l. of 30% sugar solution. To one batch add 100ppm SO_2 as sodium metabisulphite. Install the glass sheathed probe in one of the bottles as near to the point of slowest heating as possible. Use bent pipe-cleaners to keep it in place. Prepare sufficient peeled and sliced apples to fill seven jars and fill them, including the one with the probe. Add the syrup to them, filling to within 1 cm of the lip, put on the rubber seal, glass lid and screw the plastic ring in place, finally unscrewing it ⅛th turn to allow the steam to escape. Place all the bottles up to their shoulders in cold water and set up the computer to record temperature and F. The program is set to record F values relative to 85°C.

Heat the water slowly to 95°C and maintain the temperature monitoring constantly the accumulating F values in the bottom left of the screen. Remove two bottles at $F = 0.3$, two at $F = 0.6$ and two at $F = 1.2$. Prepare more apples and repeat this with the other syrup. Remember to start with cold syrup in the monitoring bottle.

Store the samples at room temperature and check them two or three times a week for the first sign of mould growth.

MAGNETISM, ELECTRICITY AND ELECTRONICS

The Hall effect comes of age

P.H. BLIGH, R. HAYWOOD AND J.J. JOHNSON
School of Applied Physics, Kingston Polytechnic, Surrey

Abstract: Appropriate electronics and software have been developed to demonstrate how a physical effect can be used as the basis for a modern instrument with its complete control of physical measurement and tailor-made computer analysis and presentation of data. The simplest and cheapest components have been used. The experiment is suitable as an introduction to instrumentation for sixth form or first year undergraduate physical science.

The Hall effect is a bit of vintage physics. Like the Stirling engine it is part of our heritage from the 19th century which plundered so many of the great mysteries of science and sampled some of its best effects.

But to young people brought up on TV documentaries, yesterday's science has the musty air of school laboratories and museum equipment. We try and make our physics courses more palatable and up-to-date by adding spoonfuls of software technology and microelectronics: the really clever trick is to be able to show the latter is a progeny of the former.

The Hall effect affords such an opportunity in that it has experienced a kind of renaissance thanks to the microtechnology revolution. The effect need not therefore be taught as a piece of past history but as a phenomenon at the heart of our technology which has also helped to make modern physics.

It was in 1879 at John Hopkins University that E.H. Hall first discovered that a magnetic field placed across a current-carrying conductor can generate a voltage across the conductor. It was not until early in the 20th century that H.A. Lorentz and Paul Drude developed theories of conduction that were able to explain the effect. In particular the ratio $V_H t/I_A B$ is a constant which is a property of the material carrying the current and called the *Hall coefficient*, R_H. V_H is the Hall voltage and t the thickness of the material carrying the current I_A with a magnetic field B acting across the conductor. R_H equals the reciprocal of the *charge carrier density* times the electronic charge. The *mobility* of the carriers equals the ratio of R_H to the resistivity of the material.[1] Small wonder the effect was used to determine these essential electrical properties of materials and classify them according to their electrical properties. The baffling results that ensued for materials we now classify as semiconductors found clarification with the arrival of quantum mechanics in the 1920s which alone was able to explain their peculiar behaviour.[2]

Not only is the effect currently important in the analysis of semiconductor materials: Hall effect sensors are big business. The Hall voltage depends both on the *current* in the conductor and the *magnetic field* across it. In these respects it is presently finding applications both as a *switch* in *control systems* and in current and flux *measurement* and *instrumentation*. HEME International, for example, a member of the Pilkington Electro Optics Division, gets its name from the fact that it produces Hall Effect Monitoring Equipment, in particular clip on Hall effect ammeters which can measure currents in conductors up to 1000 A to within 1%. They can handle all kinds of current measuring jobs non-invasively, both a.c. and d.c., embracing variable speed drives, motor vehicle electrics, welding and electroplating.[3] As a switch the Hall effect can form the basis of a smart controller to improve electronic ignition performance, not only to eliminate troublesome breaker points but help to maintain maximum spark energy and low dissipation under difficult conditions, such as rapid acceleration and cold starting.[4] Devices such as this have had to wait for the arrival of integrated circuits to come of age. Before this, such devices were complex, expensive and plagued with noise problems and temperature variation. In contrast, Hall effect ICs can combine voltage generators, signal amplifiers, Schmitt trigger circuits on a single IC chip making them inexpensive, immune to noise and temperature stable.[2]

Any experimental work in the teaching laboratory *must* reflect the fact that the Hall effect has come of age with the coming of the integrated circuit and the role it now plays in instrumentation and control. The experimental set up used here is designed to demonstrate complete control of physical measurement and analysis by computer software which is the basis of modern instrumentation.

We have used the simplest and cheapest components most available to

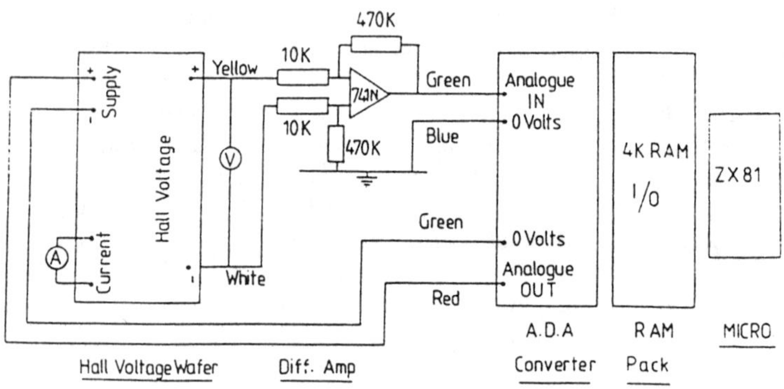

FIG. 1. Diagrammatic representation of the experimental system.

school physics. The Hall effect apparatus is that marketed by Unilab and a ZX81 is used to control its operation via a D/A, A/D interface (ADA Converter) marketed by Griffin.

Figure 1 is a diagrammatic presentation of the system and its associated hardware. The reason for using an operational amplifier to read the Hall voltage is twofold:
(a) to overcome the problem of earthing in that the ADA converter has a common earth line which, if included in the Hall voltage measuring circuit, would link it to the input circuit;
(b) to amplify the Hall voltage (V_H) to match the range of input voltages of the A/D converter.

Using the D/A to provide a range of applied voltages from 0 to 1.5 volts requires an amplification of about 50 V_H.

Figure 2 gives a print out of the software used to drive the hardware. The *software* is written to perform five function. It is designed:
(a) to *calibrate* the A/D input (IN) and D/A output (OUT) voltages in terms of the corresponding Hall voltage (V_H) and input current (I_A) which they represent. (It is the linear relationship between these two parameters that we wish to demonstrate and use to calculate the Hall coefficient R_H.)
(b) to *generate* a sequence of values of I_A and record the corresponding values of V_H.
(c) to *display graphically* the relation between I_A and V_H and *calculate* R_H from these values using a least-squares fit routine.
(d) to provide an *interactive sequence* by which the electrical parameters, resistivity, carrier density and mobility may be obtained and displayed.
(e) to use the apparatus now calibrated as an *instrument* to perform as (i) an *ammeter*, and (ii) a *Hall probe* to measure and display in real time the strength of magnetic fields in which the semiconductor chip is places.

FIG. 2. Print out of software for ZX81 computer.

Title

```
 10 REM "HALL"
 20 PRINT "* HALL EFFECT ******
"
 30 PRINT "* KINGSTON POLYTECHN
IC **"
 40 PRINT "* NOVEMBER 1984 ****
**"
 45 PAUSE 150
 50 CLS
```

Initialise address of port

```
60 LET X=22500
```

Send voltage to wafer

```
62 POKE X,120
```

Introduction

```
  60 LET X=22500
  62 POKE X,120
  70 GOSUB 2000
  80 PRINT AT 3,5;"HALL EFFECT"
  90 PRINT
  95 PRINT  THIS PROGRAM WILL.
 100 PRINT
 110 PRINT TAB 5;"1. CALIBRATE T
HE COMPUTER."
 115 PRINT TAB 5;"2. MEASURE THE
 HALL COEFFICIENT."
 120 PRINT TAB 8;"NO. OF CARRIER
S."
 130 PRINT TAB 8;"RESISTIVITY."
 140 PRINT TAB 8;"MOBILITY OF CA
RRIERS"
 150 PRINT TAB 5;"FOR YOUR SAMPL
E."
 160 PRINT TAB 5;"3. USE THE SAM
PLE TO MEASURE THE FIELD STRENGH
 OF ANY MAGNET."
 161 PRINT
 165 GOSUB 2000
 170 GOSUB 5000
```

Calibration

```
 180 CLS
 190 PRINT "CALIBRATION OF THE C
OMPUTER."
 200 GOSUB 2000
 210 PRINT
 220 PRINT "PLEASE TYPE IN:"
 230 PRINT
 235 PRINT "FIELD STRENGH (TESLA
):";
 240 INPUT N
 250 PRINT N
 260 PRINT "THICKNESS (MM):";
 270 INPUT T
 280 PRINT T
 290 PRINT "WIDTH (MM):";
 300 INPUT W
 310 PRINT W
 320 PRINT "LENGH (MM):";
 330 INPUT L
 340 PRINT L
 350 PRINT "MILLIAMPS ON INPUT:"

 355 INPUT M
 360 PRINT M
 365 PRINT "HALL VOLTAGE (MV):",
 370 INPUT V
 380 PRINT V
 385 PRINT
 390 GOSUB 2000
 400 PRINT AT 20,0;"ALL DATA COR
RECT? Y/N"
 410 IF INKEY$="Y" THEN GOTO 420
 411 IF INKEY$="N" THEN GOTO 180
 412 GOTO 400
 420 GOSUB 5000
```

Plots graph

```
 430 LET C=PEEK X
 440 LET R=M/120
 450 LET S=V/C
 460 CLS
 470 GOSUB 4000
 480 LET D=0
 485 LET E=0
 490 LET F=0
 495 LET G=0
 500 FOR A=0 TO 11
 510 POKE X,15*A
 520 LET B=PEEK X
 530 LET D=D+15*A*R
 540 LET E=E+B*S
 550 LET F=F+15*A*R*B*S
 560 LET G=G+(15*A*R)**2
```

```
570 PLOT 5*A,(B/6)+2
580 NEXT A
590 GOSUB 5000
600 CLS
```

Calculates parameters of wafer

```
610 GOSUB 2000
620 PRINT AT 3,5;"RESULTS."
630 PRINT
640 PRINT "SLOPE OF GRAPH= ";
650 LET K=(F-(D*E/12))/(G-(D**2
)/12)
660 GOSUB 3000
665 PRINT
670 PRINT "HALL COEFFICIENT= ";
680 LET K=((K*T)/(1000*N))
690 GOSUB 3000
695 PRINT
700 LET P=K
715 PRINT
720 PRINT "NO. OF CARRIERS= ";
730 LET K=1/P*1.602
740 GOSUB 3000
750 PRINT " X 10+19 PER M.CUBED
"
755 PRINT
760 PRINT "RESISTIVITY= ";
770 LET K=(W*T)/(L*R*100)
780 GOSUB 3000
790 PRINT " OHMS/METRE."
800 PRINT
804 PRINT "MOBILITY OF CARRIERS
= ";
805 LET K=P/K
806 GOSUB 3000
808 PRINT " M.SQUARED PER VOLT.
SEC"
809 PRINT
810 GOSUB 2000
820 GOSUB 5000
830 CLS
```

Uses wafer as Hall probe

```
840 PRINT "DO YOU WISH TO MEASU
RE THE FIELD STRENGH OF ANOTHER
MAGNET (USING THE APPARATUS AS A
 HALL PROBE)? Y/N"
850 INPUT A$
860 IF A$="Y" THEN GOTO 900
870 IF A$="N" THEN GOTO 1500
880 GOTO 830
900 CLS
910 PRINT "PLACE APPARATUS IN T
HE FIELD OF THE NEW MAGNET"
920 PRINT "PRESS S AND HOLD TO
STOP THE MEASUREMENTS."
930 PRINT "PRESS RETURN TO STAR
T"
940 INPUT A$
```

Makes measurements of field every second

```
950 PRINT
960 GOSUB 2000
965 PRINT
966 PRINT
970 PRINT TAB 2;"MAGNETIC FIELD
 STRENGH= "
980 PRINT AT 18,2;"WEBERS PER M
ETRE SQUARED"
990 PRINT
1000 GOSUB 2000
1010 POKE X,120
1020 LET K=(T*PEEK X*S)/(M*P*100
0)
```

```
1030 PRINT AT 14,8;"";
1040 GOSUB 3000
1050 PAUSE 100
1060 IF INKEY$="S" THEN GOTO 140
0
1070 GOTO 1020
```

End of program

```
1400 GOSUB 5000
1500 CLS
1510 PRINT "READY"
1520 STOP
```

Subroutine produces stars

```
2000 REM *STAR*
2010 PRINT "*********************
**********"
2020 RETURN
```

Subroutine reduces K to three decimal places

```
3000 REM *3 DEC PLACES*
3010 LET K=(INT (K*1000))/1000
3020 PRINT K;
3030 RETURN
```

Subroutine produces the graph axis

```
4000 REM *GRAPH AXIS*
4010 FOR Y=0 TO 63
4020 PLOT Y,2
4030 NEXT Y
4040 FOR Z=2 TO 43
4050 PLOT 0,Z
4060 NEXT Z
4070 PRINT AT 2,2;"IX / V.HALL"
4080 RETURN
```

Subroutine produces a copy if necessary

```
5000 REM *COPY*
5010 PRINT AT 21,0,"C FOR COPY,R
ETURN TO CONTINUE"
5020 INPUT A$
5030 IF A$<      THEN GOTO 5050
5040 COPY
5050 RETURN
```

Because the experimental procedure is now trivial, the student must be interrogated to gain maximum benefit from the experience. Possible student responses are as follows:

1. What are the roles of the A/D and D/A converters?

2. Give two reasons why a differential amplifier must be used to measure the Hall voltage.
3. LIST the program and write down the line numbers which cover each of the separate operations of the software.
4. Write down the values of R_H, carrier density, sign and mobility and resistivity of material as displayed. Compare these with published values for the material (or those obtained by manual operation).
5. Place the Hall chip in various well defined fields of known strength and compare these with values given by the Hall probe.
6. Draw a *flow diagram* of the system as an instrument for measuring magnetic field strength.
7. Outline the *design* features of a Hall effect control system to stabilise the current from a transformer at 100 A.

Note: Since this chapter was written, the ZX81 microcomputer has been removed from the commercial market. The experiment can be run just as easily on the Sinclair ZX Spectrum 16K or 48K machines. Motherboards for ZX81 and Spectrum machines and AD and DA converters are obtainable from Velleman (UK) Ltd, PO Box 30, St Leonards on Sea, East Sussex, and Eprom Services, 3 Wedgewood Drive, Leeds. To construct a simple ADC and DAC unit, refer to Radio Spares data sheets nos. 4642 and 5291. These circuits will need to be calibrated. Note that when using peripherals other than those described in the experiment, changes will need to be made to the program; the output address X = 22500 may need to be changed (LINE 60).[5]

REFERENCES

1 B.G. Streatman: *Solid State Devices*. Prentice Hall (1972).
2 S. Emary: Hall effect sensors — history and applications, *Transducer Technology* (March 1985).
3 F. Hutchinson: Hall effect ammeters provide accurate and versatile monitoring, *Electronics Industry* (April 1985).
4 A. Milazzotto: Smart controller improves electronic ignition performance, *Electronics Industry* (April 1985).
5 P.H. Bligh, J.J. Johnson and J.M. Wood: Automating the Hall effect, *Phys. Ed.* **20**, 246–247 (1985).

Magnetic multipoles

D.G. SMITH

Department of Physics, University College of North Wales, Bangor, LL57 2UW

Abstract: A magnetic multipole apparatus for the physics teaching laboratory is described. The apparatus enables the student to measure the magnetic field configurations of different coil arrangements, and hence to derive magnetic multipole moments. The study of multipole components provides insight into field and source symmetries, and consolidates ideas common to gravity, electrostatics and magnetostatics.

In introductory electricity and magnetism, students meet electric and magnetic dipoles and learn the significance of the concept of the dipole moment of an electric or magnetic source. Because of the nomenclature used, it is not surprising that some students get the erroneous idea that the point-charge pair system in electrostatics and the simple current loop system in magnetostatics each possesses only a dipole moment. In principle, the field configuration of each system requires an infinite set of multipole moments for its characterisation.

Although the multipole expansion of potential can be used for any field that satisfies Laplace's equation, one's choice of field for an experimental investigation is restricted by the problems of field measurement. In the case of the gravitational field it is certainly possible to investigate the multipole moments of the earth indirectly through the study of the orbits of artificial satellites. However, a laboratory-scale study of, for example, the multipole moments of a non-spherical mass of lead on the lines of Cavendish's method for '*G*' would be extremely difficult to carry out. In the case of the electric field a study of the field configurations of charged objects of differing shape should be possible, although the effects of stray charges in the vicinity of the apparatus might represent a severe limitation to measurement accuracy. The use of the magnetostatic field for such an investigation is relatively straightforward. The apparatus described here allows the magnetic multipole field components of a system of one or more current loops to be investigated, and for the values of the magnetic multipole moments of the system to be derived.

The basic experimental arrangement is shown in Fig. 1. The magnetic field strength **B** at the point P (r,θ) is given by $\mathbf{B} = -\nabla S$, where S is the magnetic scalar potential at P. The multipole expansion of S gives

$$S = (\mu_o/4\pi r)[M_o + (M_1/r)\cos\theta + (M_2/r^2)(3\cos^2\theta - 1)/2 + \dots (M_l/r^l)P_l(\cos\theta) \dots]$$

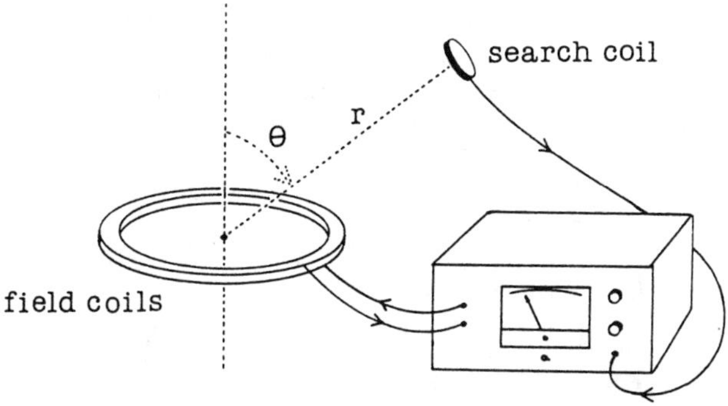

FIG. 1. Experimental arrangement.

where the M_l are the magnetic multipole moments of the field source about the origin, and the $P_l(\cos\theta)$ are the Legendre polynomials in $\cos\theta$. The choice of the origin is arbitrary, but the values of the M_l will in general, but not always, depend on the location of the origin. The apparatus is shown in block form in Fig. 2. The field coils carry a current that commutates at about 70 Hz. The signal induced in the search coil is amplified and detected using a phase sensitive detector that takes its reference from the field coil current. The circuit is described in detail in Ref. 1.

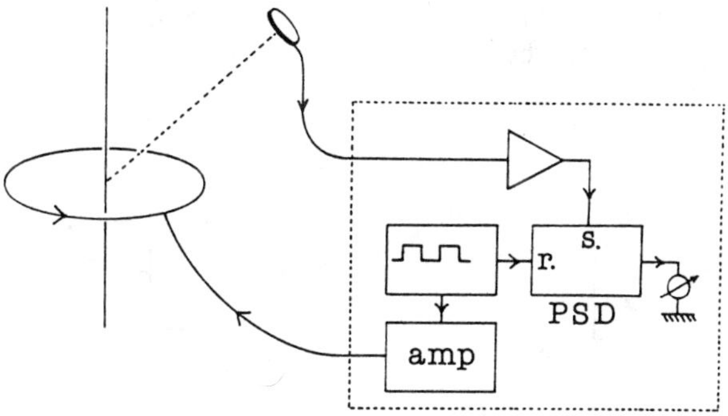

FIG. 2. Circuit of apparatus.

The apparatus may be used with different field coils. For a single circular coil of radius a, carrying total current i and with the origin at its centre, $M_o = 0$, as always (unless there are magnetic monopoles present!). The next few moments are as follows:

$$M_1 = \pi a^2 i$$

$$M_2 = 0$$

$$M_3 = -\tfrac{3}{4}a^2(\pi a^2 i)$$

$$M_4 = 0$$

$$M_5 = \tfrac{5}{8}a^4(\pi a^2 \mathrm{i})$$

$$M_6 = 0$$

A suitable coil could have a = 250 mm and about 200 turns. To what extent a given moment is significant will depend on how near the field point P is to the coil. Although it is convenient if the radial and polar field components, $B_r \equiv -\delta S/\delta r$ and $B_\theta \equiv -(1/r)\delta S/\delta\theta$ respectively, are measured by the student, it is not essential that these components be used. Likewise, there may be some merit in leaving the student to decide how she is going to measure the field point location. It will probably be found that different students will use different arrangements of metre sticks, protractors and plumb lines.

For sources of more than one coil the value of M_l for the coil system about a given origin is obtained by summing the M_l values of each of the constituent coils about that origin. It is possible to suppress by cancellation certain moments this way. The coil system shown in Fig. 3(a) has the odd-l moments about the origin (at the centre of the system) suppressed.

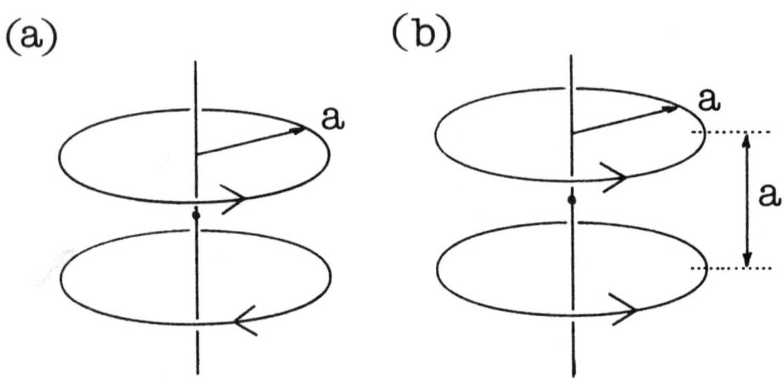

FIG. 3. Two 2-coil systems.

The field configuration now has components only for $l = 2,4,6,8$ ....., and because the $l = 2$ field component is dominant the system is termed a magnetic quadrupole. If the coil separation has the particular value of $\sqrt{3}a$, then $M_4 = 0$ also, and the $l = 2$ field component becomes relatively even more dominant. On the other hand, the coil system shown in Fig. 3(b) has all the even-l moments about the origin suppressed, and in addition, by virtue of a coil separation value of a, has $M_3 = 0$. Thus a Helmholtz coil system has zero octupole moment about its centre, and as a consequence is a better source of dipole field than is a single circular coil. For measurements on systems of this kind, the apparatus includes a set of four coils, each of mean radius 50 mm and having 500 turns.

The use of the apparatus can vary according to the needs of the student and the priorities of the laboratory class supervisor. For example, the emphasis could be on vector fields and the solution of Laplace's equation, or on the Legendre polynomials and the geometrical symmetries of field sources. In general, a laboratory-based study of multipole fields allows important ideas common to gravity, electrostatics and magnetostatics to be exercised and consolidated.

REFERENCES

1 D.G. Smith: *Am. J. Phys.* **48,** 739 (1980).

An electrical analogue of steady state thermal conduction

J.I. MISSEN
formerly of Poynton Laboratory, The City University, London

Abstract: Conduction of electricity between electrodes in an electrolyte is an analogy of steady state heat conduction using the same geometry, since both problems can in principle be solved by application of Laplace's equation. It is thus possible to solve a number of thermal problems by use of an electrolytic tank. In order to prevent electrolysis and consequent corrosion of the electrodes, the electrical supply should be low frequency a.c. (several hundred hertz).

1. Heat flow between co-axial cylinders

The heat flow q in Joules Sec^{-1} (watts) between two long co-axial cylinders of radii r_1 and r_2 and length l separated by a medium of thermal conductivity k (W m^{-1} °C^{-1}) is

$$q = [2\pi kl(T_1 - T_2)]/[\ln (r_2/r_1)] \tag{1}$$

T_1 and T_2 are the temperatures of the inner and outer cylinders respectively.

If now two co-axial metal ring electrodes of radii r_1 and r_2 are immersed in an electrolyte of resistivity ρ to a depth t, then on reference to Fig. 1, for co-axial region of radius r width dr, the resistance is dR.

$$dR = \rho dr/2\pi rt$$

Thus the total resistance between the electrodes is

$$R = \rho/2\pi t \int_{r_1}^{r_2} dr/r = (\rho/2\pi t) \ln (r_2/r_1)$$

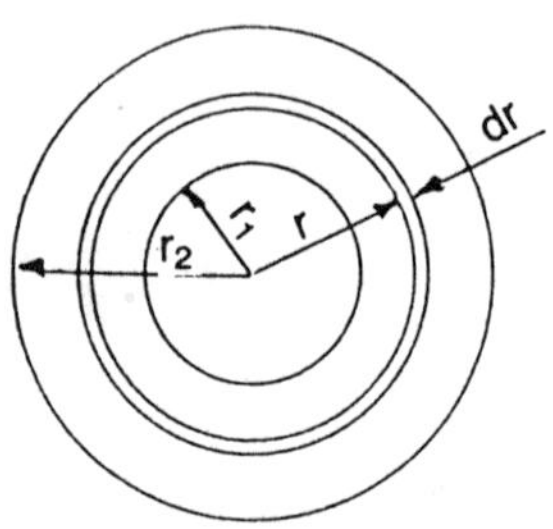

FIG. 1.

If now an a.c. potential difference $(V_1 - V_2)$ is applied across the electrodes the current flow will be

$$I = 2\pi t/\rho\ [(V - V_2)/(\ln (r_2/r_1))] \tag{2}$$

Comparison of equations (1) and (2) indicates that the following quantities must be taken as analogues for comparison.

Electrical	*Thermal*
I	q
$(V_1 - V_2)$	$(T_1 - T_2)$
t/ρ	kl

If in the thermal problem, the somwhat artificial outer boundary condition of constant temperature T_2 is replaced by natural convection to an ambient temperature T_a then

$$q = [2kl(T_1 - T_a)]/[\ln (r_2/r_1) + k/r_2h]$$

where h (W m^{-2} °C^{-1}) is the convective heat transfer coefficient.

Convective thermal resistance $R_T = 1/hA$, where A is the surface area. Thus as r_2 increases the convective thermal resistance R_T will decrease due to increasing surface area A. However the conductive thermal resistance will increase with increasing thermal path. Thus an 'optimum' value of r_2 will exist at which heat flow is maximum. By differentiation of equation (3) with respect to r_2, this can be shown to occur when

$$r_2 = k/h \tag{4}$$

For the electrical analogy, the convection can be represented by a resistance R_s between the outer cylinder and battery, as in Fig. 2.

The value of R_s can be calculated as follows:

Since $Rfl_T = 1/hA = 1/2\pi rth$ (5)

we can assume that $R_{s1} = KR_T$ (K = Constant) (6)

Thus for radius r_1, Rd2s1 $= K/2\pi r_1 th$

$$K = 2\pi r_1 th R_{s1}$$

Thus for radius r_2, $R_{s2} = (2\pi r_1 th R_{s1})/(2\pi r_2 th)$

$$\therefore\ R_{s2} = R_{s1}\ (r_1/r_2)$$

Thus for the electrical analogue

$$I = [(V_1 - V_2)]/[(\rho/2\pi t) \ln (r_2/r_1) + R_{s1}\ (r_1/r_2) \tag{7}$$

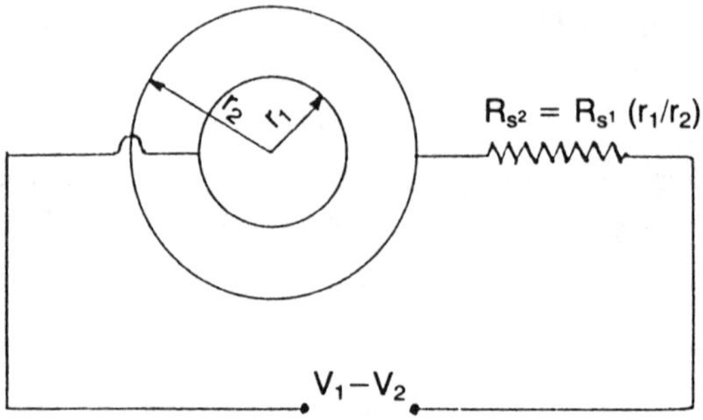

FIG. 2.

By differentiating (7) w.r.t. r_2, maximum current occurs when

$$(r_2/r_1) = R_{s1} (2\pi t/\rho) \tag{8}$$

2. Experimental procedure

Fill the dish to a depth (t) of approximately 10 mm with tap water. The dish should be level to ensure a uniform depth of electrolyte around the electrodes. Connect the 5Ω output terminals of the signal generator to a pair of circular electrodes with a milliameter in series.

With the frequency set at approximately 500 Hz adjust the output voltage to give a current of approximately 15 mA r.m.s. and measure the voltage across the plates. Repeat the experiment for 15 combinations of radii; maintain the same current I in each case.

Now from equation (2), the voltage across the plates is

$$(V_1 - V_1) = (\rho I/2\pi t) \ln (r_2/r_1)$$

By plotting $(V_1 - V_2)$ against $\ln (r_2/r_1)$, a straight line results, showing the correctness of the formula. From the slope obtain the resistivity of the electrolyte and the quantity $\rho/2\pi t$.

If now the apparatus is set up as Fig. 2 to simulate convection at the outer cylinder, plot I for the smallest central ring r_1 and a range of outer rings r_2 for constant voltage $(V_1 - V_2)$. The resistance takes values $R_{s1}(r_1/r_2)$, a suggested value for $R_{s1} = \rho/\pi t$. The current should show a maximum for $(r_2/r_1) = 2$, corresponding to optimum heat transfer.

2. Laplace's equation

Consider a long bar of triangular section with temperature T_B, T_C and T_D at corresponding sides. To calculate the temperature at any point inside the bar it is necessary to apply Laplace's equation in two dimensions.

$$\partial^2 T/\partial x^2 + \partial^2 T/\partial y^2 = 0 \tag{9}$$

On utilising a finite difference equation for the analytic expression of equation (9) using Fig. 3.

$$T_{M+1} + T_{M-1} + T_{P+1} + T_{P-1} - 4T_M P = 0$$

On applying the finite difference net to node 1 in Fig. 3

$$T_B + T_2 + T_5 + T_C - 4T_1 = O \tag{10}$$

etc.

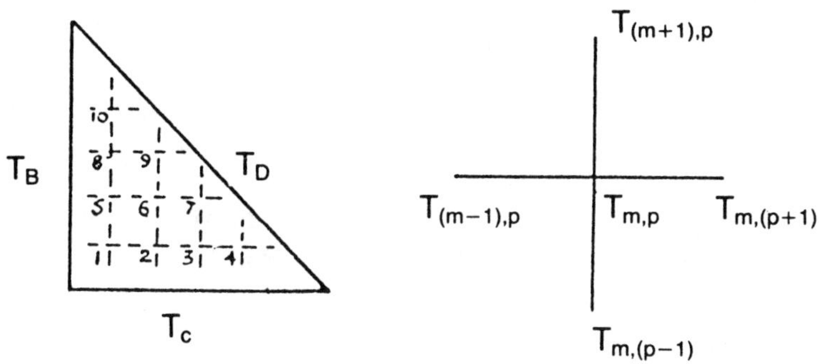

FIG. 3.

If an equation is now written for each node, the resulting 10 equations can be solved by relaxation or computer programme.

The results can be checked by the electrolytic tank and should be in close agreement.

4. Experimental procedure for Laplace's equation

Set up the triangular electrode system in the electrolyte tray and apply a.c. as shown in Fig. 4.

Measure the a.c. voltage with respect to 0 at each probe point using the a.c. voltmeter. Compare the results with those calculated from the finite difference equations.

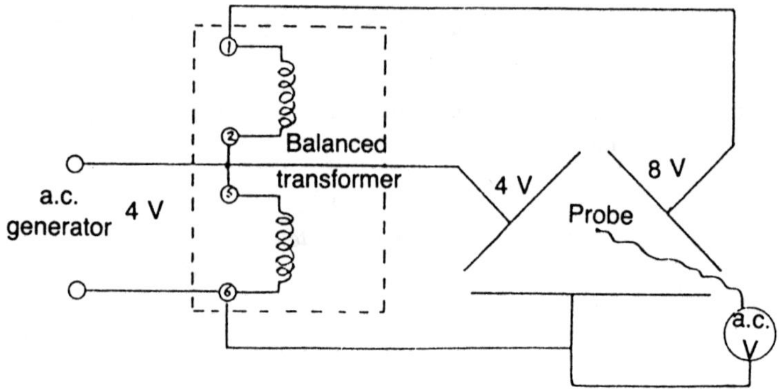

FIG. 4.

Identification of an electrodynamic force transducer (vibrator)

F. ABDULLAH
Poynton Laboratory, The City University, London EC1

Abstract: The transducer can be represented by an idealised two port network; one port being electrical the other mechanical. The measurements identify the elelctrical and mechanical parameters of the system via the electrical port.

Apparatus required

Vibrator (Ling Dynamic System Limited) in stand and with clamping arrangement; frequency oscillator (Pye Ling); two bench AVO meters; a storage oscilloscope; a scale to weigh up to approximately 100 g.

Principle of operation and simple mathematical model

The vibrator operates on the motor principle expressed by the dimple equation

$$f_m = Bli \tag{1}$$

where f_m is the force produced on a conductor of length l carrying a current i and moving within a magnetic field of flux density B. The conductor is assumed to be at right angles to the field and the resulting force is then at right angles to both the current and field, the direction being defined by the vector product **i**×**B**.

In a vibrator a uniform radial field is produced by an appropriate magnetic circuit and a coil is placed within this field. The entire length of the coil is then at right angles to the field and when the coil is excited with a current vertical motion ensues (Fig. 1).

With a sinusoidally varying current a corresponding sinusoidal force will be produced. The vibrator can operate in the audio range from 50 Hz to 5 kHz. It is principally used for vibration testing of components.

A simplified mathematical model incorporating electrical and mechanical properties of the vibrator can be expressed as a two port matrix:–

$$\begin{bmatrix} f \\ v \end{bmatrix} = \begin{bmatrix} Z_m & -h \\ h & Z_e \end{bmatrix} \begin{bmatrix} x \\ i \end{bmatrix} \tag{2}$$

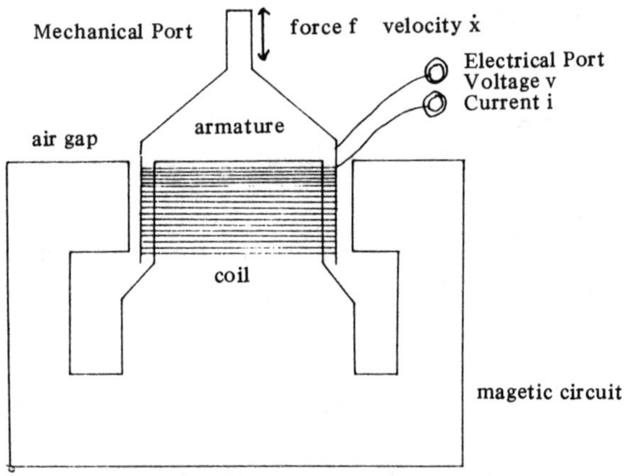

FIG. 1.

where f = force, $\dot{x}$ = velocity are mechanical port variables and v = voltage, i = current are electrical port variables, Z_m = mechanical impedance, Z_e = electrical impedance, $-h$, h are transduction coefficients. (*Note* h is identified with Bl of equation (1).)

$$Z_e = LD + R$$

where L = coil inductance and R = coil resistance

$$Z_m = mD + k/D + b$$

where m = coil assembly mass, k = stiffness, and b = damping coefficient

$D \triangleq (d/dt)$ and $(1/D) \triangleq \int dt$ are differential and integral operators respectively (t = time).

For a given vibrator it is assumed that the parameters L, R, m, k, b and h would characterise the device.

Experiments

The objective of the experiments is to determine the parameters L, R, m, k, b and h for the vibrator supplied.

These parameters will be determined by making various electrical port tests based on the mathematical matrix model of equation (2) and without dismantling the vibrator. Once the vibrator is thus characterised its performance under additional electrical or mechanical loadings could be determined.

The matrix equation (2) can be written as two equations

$$f = Z_m \dot{x} - hi \quad (3)$$

$$v = h\dot{x} + Z_e i \quad (4)$$

with

$$Z_m = mD + k/D + b$$

$$Z_e = LD + R$$

To determine, L, R, k and m sinusodial excitations of the form e^{jwt} are applied. Then at a steady state conditions we can replace $D \to jw \quad w = 2\pi f_r$ (f_r = frequency). Then

$$f = (jwm + k/jw + b)\, \dot{x} - hi \quad (5)$$

$$v = h\dot{x} + (R + jwL)i \quad (6)$$

where $f, \dot{x}$ v, i are now complex amplitudes (phasors). The displacement amplitude x is related to $\dot{x}$ by $\dot{x} = jwx$. The circuit arrangement is shown in Fig. 2.

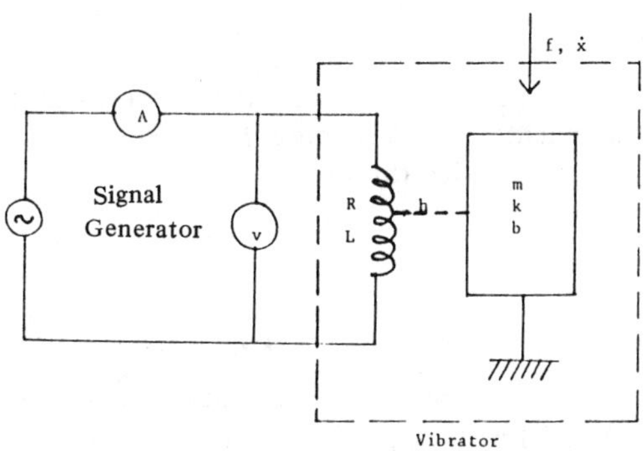

FIG. 2.

After having identified the electrical and mechanical ports of the vibrator set up the system as shown in the figure using AVO for measuring i and the other v and vary the frequency from about 60 Hz to 3 kHz setting the signal generator supply voltage to $V = 2$ Volts. At low frequencies the current will be $i = 0.6$ amps.

It is very important that you do not set the oscillator to the lowest range (× 1, 5–50HZ) as this will damage the vibrator.

Note qualitatively the change in pitch and amplitude of the vibrator as the frequency is changed.

On first switching on you may find that the vibrator does not oscillate.

This is due to stiction. Do not increase the supply beyond 2 volts; just wait for the oscillations to commence. You will be able to feel these at low frequencies by touching the bench or armature tip.

To determine R and L

Clamp the mechanical port using the clamping bar arrangement. Keep the armature tip in its free position by using the small nuts for locking. With the mechanical port firmly clamped $\dot{x} = 0$ (also $x = 0$). Then equation (6) becomes

$$v = (R + jwL)i \tag{8}$$

and therefore

$$|v/i|^2 = R^2 + w^2L^2 \tag{9}$$

Vary w and measure v, i
By plotting $|v/i|^2$ against w^2, determine R and L

To find k and m

Remove the clamping and find the mechanical resonance frequency by electrical measurements. In this situation $f = 0$ (no applied mechanical force). Then solving equation (5) for $\dot{x}$.

$$\dot{x} = hi/(jwm + k/jw + b) \tag{10}$$

and substituting in (6)

$$v = [R + jwL + h^2/(jwm + k/jw + b)]i \tag{11}$$

$hi/(jwm + k/jw + b)$ is called the motional impedance since it is due entirely to mechanical motion. [........] is the total electrical impedance.
Solving for i

$$i = v/[R + jwL + h^2/(jwm + k/jw\ b)] \tag{12}$$

Mechanical resonance occurs when $jwm + k/jw = 0$, i.e. $w^2 = k/m$. Denoting this by w_n, $w_n = (k/m)^{\frac{1}{2}}$.
At resonance (about 100 Hz) $wL << R$ (check this using values of R, L already determined) then

$$i = v/[R + h^2/(jwm + k/jw + b)] \tag{13}$$

The precise resonance can be determined by varying w and observing $|i|$ on the AVO.

At resonance $|i|$ will be a minimum, as can be seen from equation (13). Determine this resonance and the value of $|i|$ at the minimum. Then add an additional masses M by screwing on the tip of the armature and repeat the experiment. In this case resonance will occur at

$$W_n = [k/(m + M)]^{\frac{1}{2}}$$

Note This can be derived as before but now $f = -jwMx$ in equation (5).

Squaring and rearranging the above equation

$$m + M = k/w_n^{\ 2}$$

By plotting M against $w_n^{\ -2}$, m, k can be determined. Do this using the additional masses provided using a scale to weigh them.

Do you expect that the minimum of current will be dependent on M?

Having noted the current at resonance, h^2/b can be found from equation (13):

$$i = v/(R + h^2/b) \tag{14}$$

leaving us to determine either h or b.

To find b (damping coefficient)

In this case we perform a transient response test.

Switch off the signal generator and disconnect the AVOs.

Apply a mechanical step by depressing the armature tip with the tip of your nail and releasing quickly. Observe the decay of the induced waveform on a storage scope. The storage scope is measuring on open circuit so $i = 0$. Then equation (3) and (4) read

$$f = Z_m\dot{x} \tag{15}$$

and

$$v = h\dot{x} \tag{16}$$

Eliminating x

$$Z_m v = hf \tag{17}$$

Substituting $Z_m = mD + k/D + b$ remembering D and $1/D$ are operators now

$$m\ \mathrm{d}v/\mathrm{d}t + bv + k\int v\ \mathrm{d}t = hf \tag{18}$$

Differentiate this equation

$$m\ \mathrm{d}^2v/\mathrm{d}t^2 + b\ \mathrm{d}v/\mathrm{d}t + kv = h\ \mathrm{d}f/\mathrm{d}t \tag{19}$$

A mechanical step of force f is applied. $\mathrm{d}f/\mathrm{d}t$ will then be an impulse and the voltage observed will correspond to an impulse response of the second order system of equation (19).

The response is underdamped and of the form shown in Fig. 3 with an exponential decaying envelope.

Standard second order theory enables the damping ratio ξ to be calculated from

$$\xi(1 - \xi^2)^{\frac{1}{2}} = (1/2\pi n) \ln (v_o/v_n) \quad (20)$$

In practice ξ is small (ξ<<1) so

$$\xi \simeq (1/2\pi n) \ln (v_o/v_n) \quad (21)$$

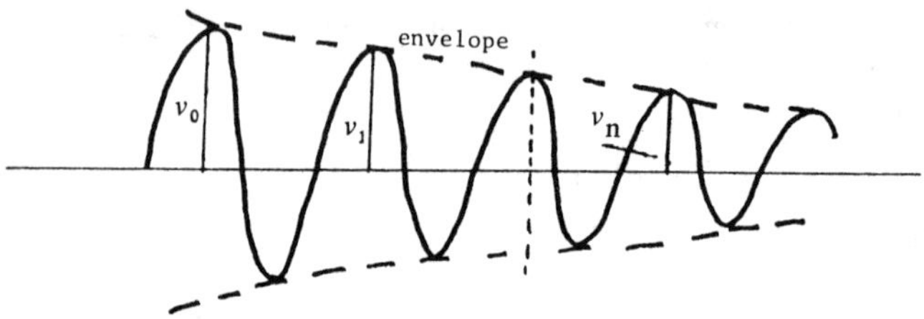

FIG. 3.

The time base can be determined from W_n (already measured). Use the internal trigger facility of the scope and the 1 volt/division voltage scale.

Determine v_o and a suitable v_n from the scope trace making a rough sketch of the trace. Hence determine ξ. By standard second order theory (see for e.g. E.O. Doeblin *Measurement Systems: Application and Design*, McGraw Hill).

$$\xi = \tfrac{1}{2}b\ (\text{km})^{-\frac{1}{2}}$$

b can now be determined and from the ratio $h^2/b,h$ can be found.

Conclusion

In your conclusion you should tabulate the results obtained for L,R,k,m,b,h and discuss sources of errors including any that may arise from your measuring instruments (e.g. bandwidth or frequency response of the AVOs). Can you think what may be the practical uses of *electrical* port testing of electromechanical machines?

Weather satellites — reception and display*

T.I. PRESTON
Physics Department, University of Leicester, University Road, Leicester LE1 7RH

Abstract: The object of the experiment is to receive information from weather satellites which use the Automatic Picture Transmission (APT) system, and to display this information, in the form of weather pictures, on a colour monitor, or a colour printer, using a BBC microcomputer. The main aim is to give the student experience in data reception, handling and interpretation, and also an insight into weather prediction.

1. Introduction

NOAA6 and NOAA7 are both low, polar orbiting earth satellites, with an orbit period of approximately 1 hour 41 minutes, during which time the earth rotates about 25 degrees.

The Automatic Picture Transmission (APT) system produces, and transmits, video and telemetry data, at the rate of 120 lines/min, of a strip of earth approximately 1900 km wide, immediately below the satellite. Each line is scanned in visible and infra red, and the forward motion of the satellite is used to increment each line. A continuous signal representing both visible and infra red is therefore produced as the satellite passes overhead. This is used to amplitude modulate on the satellites, a subcarrier wave of 2400 Hz, which in turn frequency modulates a carrier wave of 137.62 MHz (NOAA6) or 137.5 MHz (NOAA7).

Figure 1 shows the telemetry frame format. It can be seen that the maximum subcarrier modulation index of 87.0% corresponds to wedge number eight, on the telemetry grey scale. Figure 2 shows the full APT frame format, consisting of both visible and infra red video, synchronisation pulses and 128 complete lines of telemetry. Figure 3 gives details of the APT synchronisation pulses.

*This experiment was devised by Dr T.J. Harris and Dr E. Mathieson at Leicester University Physics Department, where it ran as an MSc experiment for several years, using a different method of picture reconstruction. It has been updated, for use with a BBC computer, and is now an undergraduate experiment in the Physics Department.

One complete telemetry frame

MI=10·6% Wedge 1 (1)	MI=21·5% Wedge 2 (2)	MI=32·4% Wedge 3 (3)	MI=43·3% Wedge 4 (4)	MI=54·2% Wedge 5 (5)	MI=65·2% Wedge 6 (6)	MI=76·0% Wedge 7 (7)	MI=87·0% Wedge 8 (8)
Zero modulation reference MI=0·0% (9)	Therm temp 1 (10)	Therm temp 2 (11)	Therm temp 3 (12)	Therm temp 4 (13)	Patch temp (14)	Back scan (15)	Channel ID wedge (16)

FIG. 1. Telemetry frame format consisting of 16 frames of information. The first 8 frames may be used to set up the correct signal level, when reconstructing the picture.

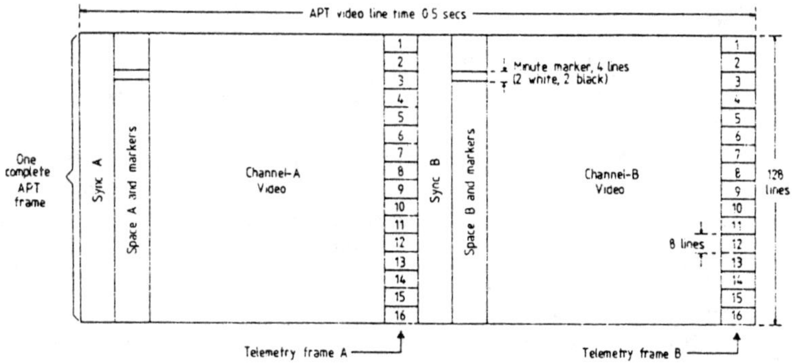

FIG. 2. One complete APT frame, showing details of the video information, marker pulses and 16 frames (128 lines) of telemetry.

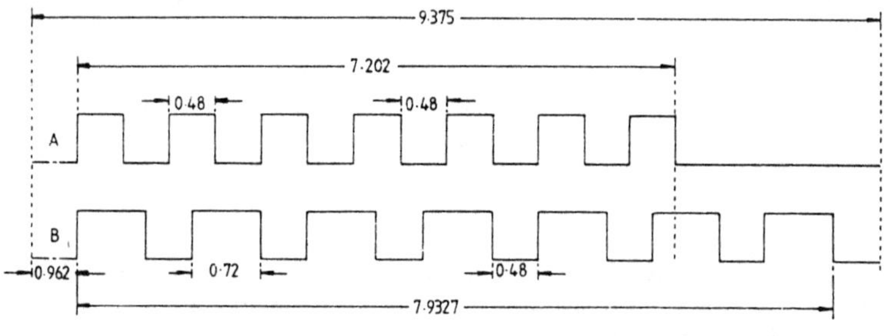

FIG. 3. Timing details of synch. A and B pulses, transmitted on each line.

2. Principle of operation

Figure 4 is a block diagram of the complete system. This may be conveniently split into two parts:

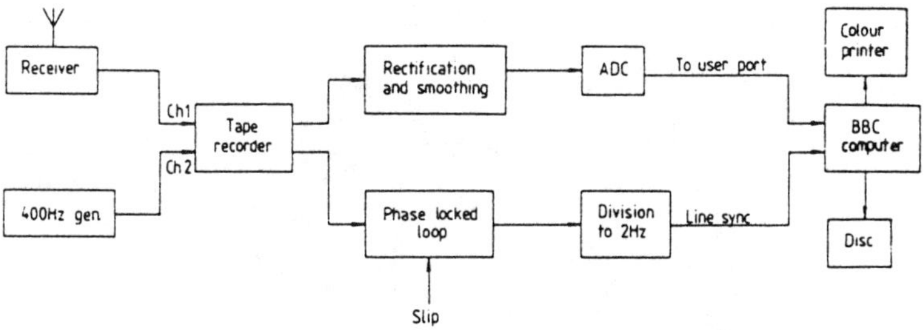

FIG. 4. Block diagram of the apparatus.

Reception

The received signal from the aerial is fed into an FM tuner, where the 137 MHz carrier is demodulated. The 2400 Hz subcarrier is then recorded on one channel of a stereo tape recorder.

We have found it better to produce our synchronisation pulses, rather than using those on the subcarrier, since these may be lost if the signal becomes noisy, or if the amplitude modulation reaches peak white level. A 400 Hz square wave is therefore generated for this purpose, and is recorded simultaneously on the second channel of the tape recorder.

Display

The recorded satellite signal is rectified and smoothed, and this analogue signal fed into an 8-bit analogue to digital converter (ADC). This digital output then goes to the user port of the BBC microcomputer, together with a pulse, derived from the 400 Hz square wave to provide synchronisation, and the picture is built up, line by line, on the screen.

Once the picture has been reconstructed on the screen, it may be stored on disc, or a screendump can be made to a colour printer.

3. The apparatus*

Figures 5 and 6 show the apparatus.

For good, non-fading reception of weather satellites, a high gain, circularly polarised aerial is required. A Yagi array gives good results but, as it

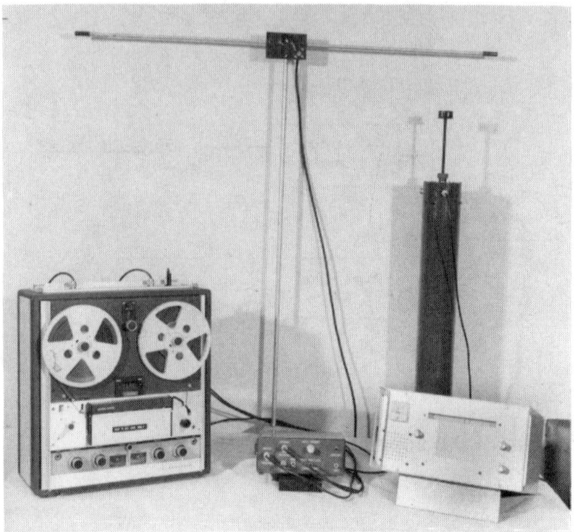

FIG. 5. Apparatus used for the reception of the satellite signal.

FIG. 6. Apparatus for reconstruction and display of the picture.

*Further details of the apparatus, including circuit diagrams, are available on request.

is highly directional, some method of steering must be used to track the satellite. This may be automatic, or manual, with the operator listening to the signal through an earphone extension, or by monitoring the level on a meter.

The simplest aerial, but one that gives reasonable results, is a horizontal dipole, or a crossed dipole, designed to operate around 137 MHz. Although the gain is not as good as a Yagi, and a certain amount of fading occurs, overhead passes may be recorded for up to 15 minutes, without moving the aerial.

The receiver consists of a commercial frequency converter, with an input sensitivity of about 1 μV, converting the signal from 136–138 MHz to 28–30 MHz, and a 'home-made' tuner with an audio output.

The synchronisation pulse generator is in two parts: a 10 MHz crystal oscillator is divided down to 400 Hz, and the output recorded on one channel of the tape recorder. The output from the tape recorder, on playback, is then fed into a phaselocked loop. By allowing the PLL to 'slip', the picture may be properly framed on the screen. The 400 Hz is then divided down further to 2 Hz, to provide synchronisation pulses to CA1 of the BBC user port.

The tape recorder used is a Ferrograph Series Seven reel to reel type, but any good stereo cassette recorder should give acceptable results.

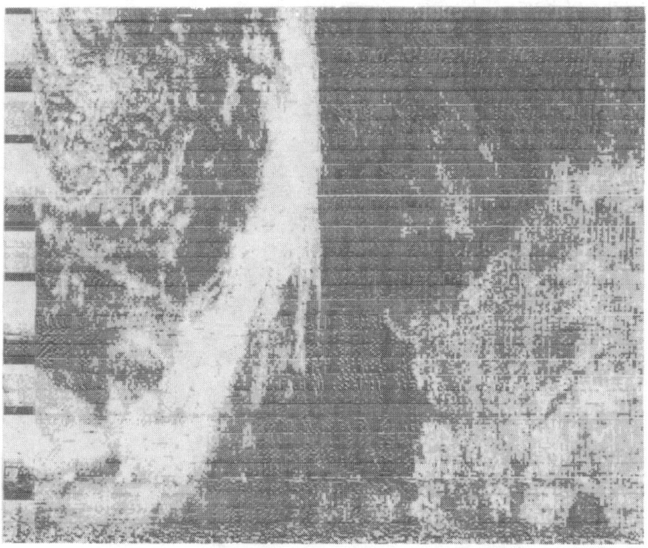

FIG. 7. Typical result from a CX80 colour printer, showing a visible weather picture. One dot printed here represents one pixel on the screen.

A CMOS operational amplifier is used to rectify and filter the signal. It also has the advantage of providing a variable amplification of the signal.

The ADC interface is based around the Ferranti 8-bit ADC. This is very fast, with a conversion time of around 9 μs. The output of this goes into the BBC user port, with CB1 and CB2 controlling the conversions.

The software† to reconstruct the picture gives a choice, in mode 1, of 320 by 256 pixels in four different colours, or in mode 2, 160 by 256 pixels with eight different colours. The picture may be built up from north to south, or from south to north, depending on whether the recording is during the day, or at night. It can then be stored on disc, if required, so that a catalogue of pictures may be created.

Another program‡ is used to produce hard copy from the screen, in mode 1, onto an Integrex CX80 colour printer.

4. Acknowledgements

I would like to thank Mr M. Furminger of Nene College, Northants, who wrote most of the software, and also Mr W.A. Johnson for his assistance with the equipment.

REFERENCES

1 Institute of Physics, *Physics Education* **6** (1971).
2 *Radio Communication* (April/May 1979).
3 *Electronics and Computing Monthly* (July 1983).

†This software, and also various commercial receivers, are available from Timestep Electronics, Wickham Brook, nr. Newmarket, Suffolk.

‡This program is available on request.

A simple pulse-height analyser

M.R. HALSE AND J. CHAN
Physics Laboratory, University of Kent, Canterbury.

Abstract: A simple multichannel analyser is described based on an Acorn Atom computer. The interface to the Atom has been kept as simple as possible and costs only a few pounds. A γ ray spectrum from a 5 milli-Curie Cs-137 source can be obtained in 4 minutes. Using machine code, a full spectrum takes 30 seconds.

A simple and inexpensive multichannel analyser makes a very useful addition to our second year experiment on γ ray spectroscopy. A γ spectrum from a 5 milli-Curie Cs-137 source can be obtained in a matter of minutes with the minimum of additional equipment. The student thus gets a good qualitative feel for the experiment before proceeding to more detailed measurements using a single channel analyser.

The electronic interface was designed by one of us (JC) as part of an MSc project. No attempt has been made at sophistication in the electronics; all the ICs are readily available and the total cost should be no more than a few pounds. The Acorn Atom computer was chosen as it has the necessary 6522 VIA built in. The Atom also has quite adequate graphics for our purposes. It was felt important not to tie up a BBC computer permanently on an experiment used only twice a week for only part of the year.

Experimental arrangement (Fig. 1)

Incident γ rays interact with a thallium doped NaI crystal scintillator to produce photons. Each γ ray produces many optical photons. These are detected by a photomultiplier coupled to the scintillator. The output pulse from the photomultiplier is (ideally) proportional to the incident γ ray energy.

Circuit description

The pulses produced by the NaI/photomultiplier combination are typically 400 ns wide and 8 V zero to peak and thus too fast to be of use directly. An RC integrating circuit (see Fig. 2a) with a time constant of 8 μs stretches the pulse to a fraction of a millisecond, thus allowing amplification by a 741. The stretched pulse charges up a peak hold capacitor C

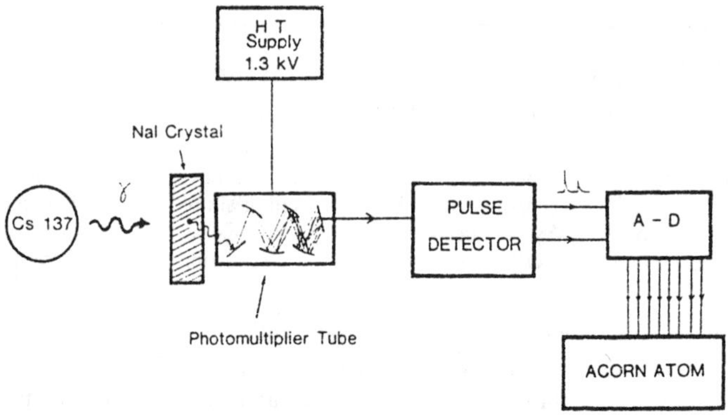

FIG. 1. Block diagram of apparatus.

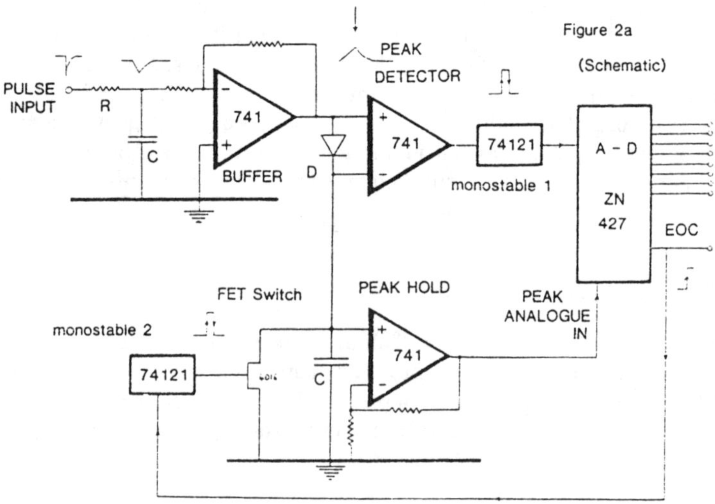

FIG. 2. (a) Simplified interface circuit diagram (schematic only).

through diode D. On the falling part of the pulse, diode D is reverse biased and capacitor C is isolated. The voltage on C thus equals the peak value of the pulse. The sudden reverse bias on diode D also causes the 741 comparator to flip, thus triggering monostable 1 and the ZN 427 A to D converter. When the peak pulse voltage has been converted, the EOC (end of conversion) signal latches the 8 bit output, informs the computer that valid data is present and discharges capacitor C through monostable 2 and the 4016 quad FET switch.

Complications arise at high count rates due to overlapping pulses. Some pulse heights are underestimated and other pulses are missed altogether. These errors cannot easily be avoided with such simple electronics. A crude remedy is to increase the source detector separation. The full circuit diagram is shown in Fig. 2b.

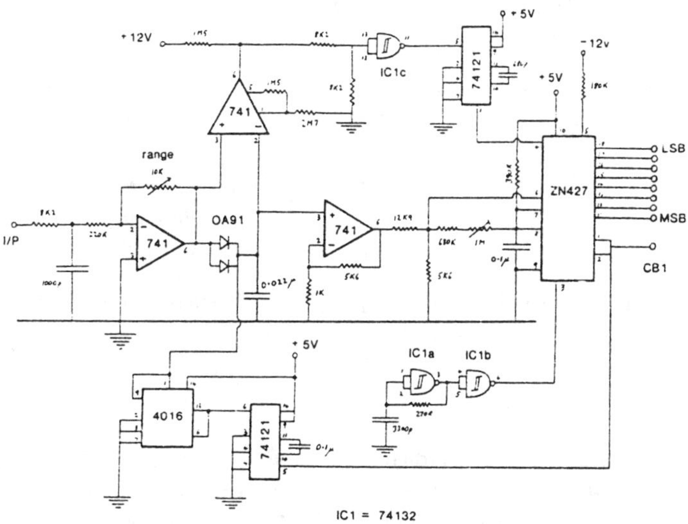

FIG. 2. (b) Full interface circuit diagram.

A ZN 428 digital to analogue converter can easily be added to the interface if a hard copy to a chart recorder is required. Alternatively, the interface could be used without modification on a BBC computer. In this case a screen dump would be the most convenient way of obtaining hard copy.

Computer program

The data supplied by the A to D converter to the Acorn Atom is a binary number representing the strength of the input pulse concerned. Each time a pulse is detected, this number directs the Atom to one of 254 memory locations and its contents are incremented by one. (Pulse heights of 0 and 255 are ignored.) As the tally of pulse heights build up, the contents of these memory locations are also duplicated in screen memory so that the growth of the energy spectrum can be seen in real time. The program (written in Atom Basic) takes approximately 4 minutes to build up to a full screen spectrum, such as shown in Fig. 3.

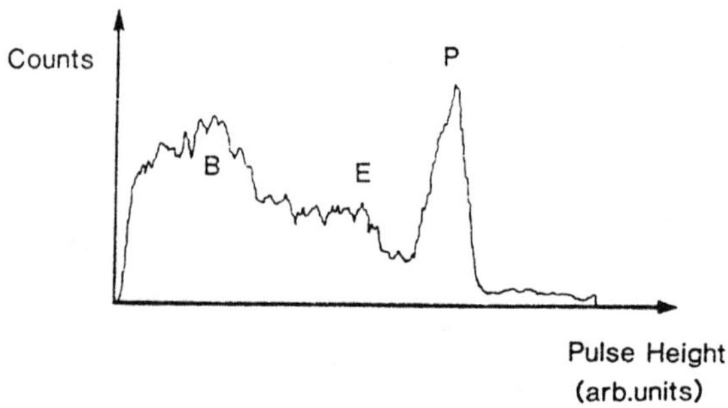

FIG. 3. Typical γ ray spectrum (Cs-137) produced by the interface of Fig. 2b and an Acorn Atom computer.

The natural background radiation can be subtracted out of the readings by re-running the program without the source. Rather than use the internal 6522 timer and interrupts for this purpose, timing can either be done manually (e.g. 4 minutes with source and 4 minutes background only) or more crudely by ensuring that the count of all low energy pulses (≤ 0.1 of the γ peak) is the same in both cases; i.e. in an energy range where the source contribution is negligible. A full listing of the program can be obtained by writing to the authors.

Results

Figure 3 shows a pen recorder trace of the γ ray spectrum of Cs-137 taken with the multichannel analyser. The γ ray peak P (0.66 MeV) is clearly visible together with the Compton edge E and backscatter region B. Because of the weakness of the particular source used, the low energy part of the spectrum contains a very considerable contribution from the natural background. This can easily be subtracted out, if required (see above).

REFERENCES

1 Eisberg: *Fundamentals of Modern Physics,* p. 82.
2 M. Cooper and D.P Woodruff: 'A Compton scattering experiment', Institute of Physics, Exhibition of Undergraduate Experiments, Imperial College (1984).
3 J.K. Jones and A.R. Clarke: 'VELA in undergraduate teaching', *ibid.*
4 J.A. Hodgkinson: 'Pulse height analysis and multiscaling using a microcomputer', *ibid.*
5 S. Chomet, M.A. Hill and D.P Lidiard: 'Beta spectroscopy for pedestrians', *ibid.*

COMPUTING

Computer interfacing experiments

P.A. Bates and E. Webster
School of Physics and Astronomy, Lancashire Polytechnic,
Preston PR1 2TQ

Abstract: Details are presented of a circuit which may be connected to the user port of a BBC Microcomputer. This device enables digital communication in and out of the computer and is used as the basis of instruction in computer interfacing. Further circuits are briefly described which form the basis of a structured introduction to computer interfacing.

Introduction

The series of experiments described below are used by HND applied physics and BSc physics students as an introduction to the use of computers in physics. The computer used is the BBC Microcomputer and all experiments interface to the user port which means that all communication is undertaken using the indirection operators. Using this technique means that by learning only a few extra BASIC commands the student can use their knowledge of programming gained from conventional computer courses.

The format of the experiments is that the student is given details of the method of communication between the device and the microcomputer and this is followed by suggestions of what experiments may be performed with it. Once the student has achieved communication with the device, he/she is then able to write a program that will either control or make measurements with suitable information presented on the computer's display screen.

Four circuits form the basis of the series of experiments. Three of the circuits, stepper motor control, analogue to digital conversion and digital to analogue conversion will be outlined whilst the user port/VIA tester will be described in detail.

The BBC user port

The BBS user port is the 'B' side of a 6522 versatile interface adapter (VIA). It consists of 8 input/output ports (B0–B7) and two control lines CB1 and CB2. A comprehensive guide to the VIA may be found in the *Advanced User Guide for the BBC Microcomputer.*[1]

In these series of experiments the student only requires to know certain BASIC statements which configure the VIA. These statements can be written in either hexadecimal or decimal notation and are as follows:

VIA as an output: ? & FE62 = &FF or ?65122 = 255
VIA as an input: ? & FE62 = &00 or ?65122 = 0
CB2 of VIA activated: ? FE6C = &CC or ?65132 = 204
CB2 of VIA deactivated: ? & FE6C = &CE or ?65132 = 206

The activation and deactivation of CB2 permits either an ADC or DAC to be switched on or off.

Reading or writing to the VIA may be performed in a similar manner using the appropriate memory location.

Read VIA : X = ? & FE60 or X = ?65120
Write to VIA : ? & FE60 = X or ?65120 = X

where X is the data.

Once these statements have been introduced to the student, he/she only needs to know the special operating requirements of the individual interface module efore he/she embarks upon the particular experiment. The student is therefore not involved in complicated setting up procedures and more time can be devoted to program presentation and actual physical measurements.

The interface modules

Each experiment is based upon an interface module, i.e. user port tester, stepping motor controller, ADC and DAC. These modules have been constructed on small printed circuit boards with a 20-way IDC connector enabling 20-way ribbon cable to connect the module to the BBC Microcomputer user port.

In addition to the data and control lines of the VIA there are +5 V and 0 V power supply lines which are adequate to power some of the interface modules which do not require external power supplies.

The user port tester

The user port tester was originally developed to test the performance of the VIA but it lends itself to being used as an introduction to digital input and output from the BBC Microcomputer. The interface module consists of ten identical circuits which are connected to the eight data and two control lines of the VIA. Figure 1 shows one of the ten circuits.

The state of the VIA is configured from the keyboard of the BBC Microcomputer. When the VIA is configured as an input, the position of the switch determines the state of the logic being fed into the VIA and this may be checked by observing the state of the LED. When the VIA is used as an output, the switch on the interface module is set so that the LED is

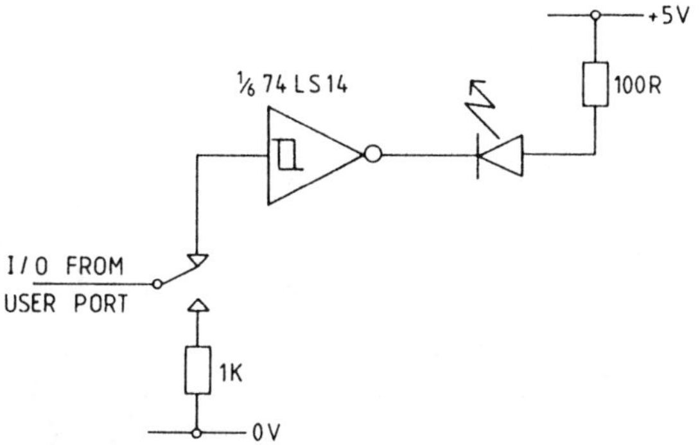

FIG. 1. User port tester circuit (each I/O of the VIA requires one of these circuits).

in circuit, thus enabling the state of the output to be immediately visible. Figures 2 and 3 show the printed circuit board and the component layout respectively. A list of components is given in Table 1. The power supply for the board is derived from the 5 V supply of the BBC Microcomputer, which is adequate in this particular case.

TABLE. 1.

	R.S. Components Cat. No.[4]
Semiconductor	
2 × 74LS14	307-547
Switch	
10 SPST DIL switch	336-680
LED display	
10 bar DIL array	588-027
Resistor	
10 × 1k	132-494
1 × 100R	132-258
Connector	
20-way IDC	469-960

The other interfaces

Once the student has become conversant with how data can be transmitted or received through the VIA, further small experiments can be developed. e.g.

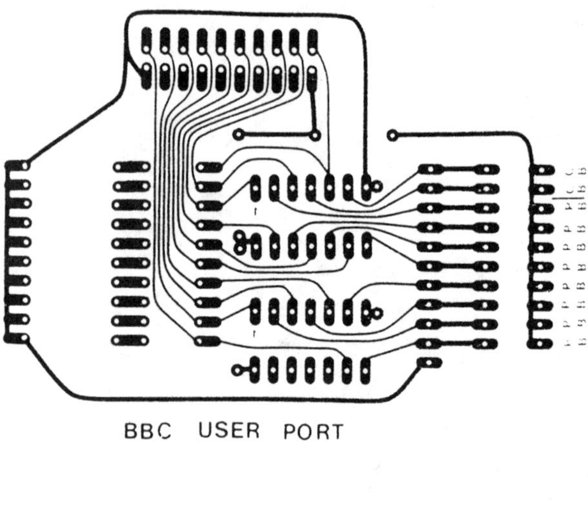

FIG. 2. User port tester printed circuit board mask.

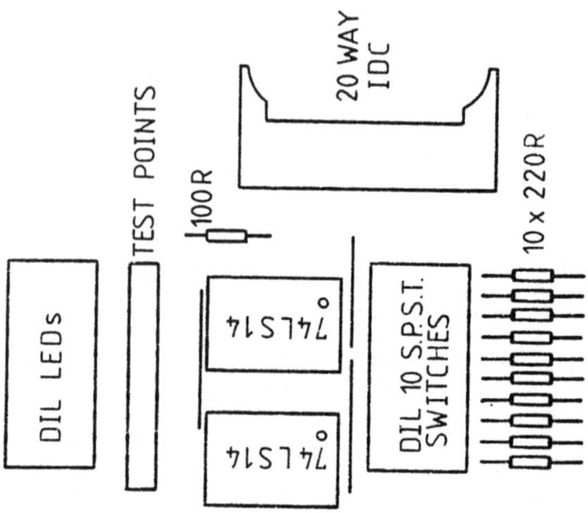

FIG. 3. Component layout for user port tester.

(i) *Stepping motor*

A small three-phase unipolar stepping motor may be controlled from the user port. Each of the coils is driven from a data line of the VIA through a Darlington driver circuit. The student has to write the correct routine which will cause the motor to rotate in the clockwise and counter-clockwise directions.

The experiment can then be developed to enable the student to write a program which causes the motor to raise or lower a small load attached to a thread wound around the spindle of the motor.

(ii) *ADC*

Unipolar, 0 V to +5 V or bipolar, −5 V to +5 V voltages may be measured using a National Semiconductors ADC0804 8-bit analogue to digital converter. The connections to the ADC are 8 data lines, +5 V and 0 V and a connection to the CB2 line of the VIA so that it may be enabled or disabled. The ADC0804 has its own internal clock and further particulars may be found in *National Semiconductors Databook*.[2] The 8-bit resolution of the ADC makes it ideal for a student to use with a thermistor and the appropriate formula to measure temperature.

(iii) *DAC*

A Ferranti ZN 428 8-bit digital to analogue converter can be used with the BBC Microcomputer user port. It can be used in either unipolar or bipolar mode and the same connections as used for the ADC are required. Further details are available in *Ferranti Semiconductors Technical Handbook*.[3]

Conclusion

All of the interface modules described above can be easily programmed with only a limited knowledge of computer programming. The student is able to concentrate upon using the interface to make measurements and thus gain confidence with such devices. This technique has proved to be helpful to students when they tackle future project work.

Acknowledgements

We wish to thank Mr L. Redpath for designing the BBC user port pcb layout, Miss H. Poulton for making the circuit boards and Mrs B. Ridding for preparing the figures.

REFERENCES

1 A.C. Bray, A.C Dickens and M.A. Holmes: *Advanced User Guide for the BBC Microcomputer*, pp. 397–415. The Cambridge Microcomputer Centre.
2 *National Semiconductors Databook, 1984,* pp. 5-12–5-43. National Semiconductor (UK) Ltd.
3 *Ferranti Semiconductors Technical Handbook*, Volume 10: *Data Converters,* pp. 1-27–1-36. Ferranti Electronics Ltd.
4 R.S. Components Catalogue, R.S. Components, Corby, Northants.

Computational physics laboratory exercises

P.H. Borcherds
Department of Physics, University of Birmingham, P.O. Box 363, Birmingham B15 2TT

Abstract: A set of exercises for an introductory laboratory course in computational physics, using BBC Microcomputers, is described. In the early exercises the emphasis is on numerical methods, in the later ones on the applications of computing in physics. The topics covered are Monte Carlo methods, fast Fourier transforms, non-linear least-squares fitting, and the restricted three-body problem.

1. Introduction

The recent and continuing developments in microcomputers, together with their continually falling prices, are transforming the way in which computers are used in physics. We believe that it is now important that these changes are reflected in our teaching. It is still too soon to make radical changes in teaching: we are too close to the advent of the computer, and too inexperienced ourselves in how to use computers for us to totally transform our teaching — anyone who doubts this need only consider some of the appalling 'educational' software available. There are some aspects of teaching involving computers which will brook no delay: we must ensure students have a modicum of competence in numerical methods, and we must also expose students to some of the applications of computers in physics.

The exercises to be described form the practical side of an optional course entitled 'Computational Physics', offered to second year physics undergraduates at the University of Birmingham.[1] There are eight weekly laboratory classes, each lasting three hours. In the first four classes the emphasis is on numerical methods, and these exercises will be described briefly. In the remaining classes there is more emphasis on physics: the topics covered are Monte Carlo methods, 'fast' Fourier transforms, non-linear least-squares fitting and the classical three-body problem.

The exercises are carried out on BBC Microcomputers, with colour displays, and all the programmes are written in BASIC.

In most of the classes the students are provided with the programmes and are required to use them, and on occasion, to assess their precision. It would be totally unrealistic to require the students to write the required programmes in the time available.

2. Numerical methods

The exercises on numerical methods make use, whenever possible, of material already familiar to the students through courses in mathematics, e.g. they are required to integrate numerically such functions as sine or cosine. For *physics* students this approach has the advantage that the emphasis is clearly on the numerical method and not on unfamiliar mathematics. In these exercises the students are required to estimate the errors and to identify their sources.

Several of the exercises are based on ones in the Cambridge Part 1B Mathematics course,[2] but have been modified to make them more suitable for physics students.

2.1 *Integration of ordinary differential equations*

In this exercise the students are required to write their own (BASIC) programmes to solve two initial value problems, namely the radioactive decay equation and the gravitational two-body problem. For each problem both the Euler method and the leapfrog method are used.

The Euler method yields an acceptable solution for the decay equation, and the leapfrog method one for the two-body problem. How each solution depends on step size is easily determined.

That the Euler method leads to an outwardly spiralling 'solution' to the two-body problem is well known, easily understood and unsurprising.

What is perhaps more surprising is the unsatisfactory 'solution' of the decay equation by the leapfrog method, since the problem is satisfactorily solved by the lower order Euler method. When 'solved' by the leapfrog method an oscillating and increasing 'solution' is obtained for the decay equation.

It is felt to be most important to emphasise and re-emphasise that a numerical method can lead to a nonsensical result. In this exercise the students find that the programme they have written leads to what they know is a nonsensical result, and they are able to explain qualitatively why that has occurred. The analytical explanation is best dealt with in a subsequent lecture.

3. Monte Carlo methods: the Ising model

The Ising model is a well known, simple, but not very good model of ferromagnetism. It does display a phase change and is easy to set up for Monte Carlo calculations.[3]

A Monte Carlo calculation is frequently used for estimating the value of a multi-dimensional integral: in statistical physics the number of dimensions is equal to six times the number of particles, and can be many thousands. To determine such an integral by regular analytically chosen sample points would require an impossible amount of effort, but an approximate value can be obtained fairly quickly using randomly chosen sample points.

In the exercise a comparatively small 'crystal' is shown on the display by colour coded atoms. The change from order to disorder as the temperature increases is clearly seen. The programme also calculates the energy and magnetisation of many configurations in a canonical ensemble, and from the variances of these quantities over the ensemble the specific heat and susceptibility are found. The discontinuities at the critical temperature are clear, although considerably smeared out as a result of the small size of the crystal and the small number of configurations.

The programme runs in BASIC, and is slow even with small crystals. In order to make a viable exercise the whole class is required to cooperate on determining the variation of specific heat with temperature and with magnetic field. This has proved to be a successful aspect of the exercise.

4. The 'fast' Fourier transform

This technique, more correctly known as the discrete Fourier transform popularised in the 1960s, is fully described by Whitaker and Robinson.[4] It is becoming widely used in experimental measurements, and is capable of extracting small signals from extremely noisy backgrounds.

In the laboratory exercises the students determine the discrete Fourier transforms of several simple functions such as a sine wave, the sum of several sine waves, and a sine wave modulated by a Gaussian. Both the original function and its transform are displayed graphically: either the magnitude of the transform or its real and imaginary components can be shown: the availability of a colour monitor is invaluable for this.

Convolutions, aliasing and leakage are demonstrated graphically. The use of windows to reduce leakage, and the consequent loss in resolution is clearly seen.

Carrying out Fourier transforms on simple functions helps the students to interpret the spectra they obtain in the laboratory: the exercises on leakage are particularly valuable in this respect.

5. Least squares fitting

Least squares fitting (also known as 'regression') is a technique much used in the laboratory to fit the 'best' straight line to a set of experimental data points. (It is now so widely used that many students do not plot a graph first.)

The technique can be extended to fitting data to non-linear functions, and in the exercises appropriate data is fitted by such functions as (1) a damped sine wave, (2) the sum of two exponentials, and (3) the sum of two Gaussians. Since the problems are non-linear it is necessary to use an iterative method.

The two simplest ways of solving such non-linear least-squares problems are either slow or unstable, that is they converge to a solution very slowly or perhaps do not converge at all. The Levenberg-Marquardt algorithm combines these two simple methods, namely linearisation and steepest descent, and does converge to a stable solution fairly rapidly.

In the exercises the data is plotted, as also is the 'model' being fitted. The improvement in the model is thus apparent to the student.

The damped sine wave data is required to be fitted by two models, a damped and undamped sine wave. A statistical F-test is then applied to find whether there is any justification for the more complicated model, i.e. whether the value of the damping constant is significantly different from zero.

This can be repeated for different data sets, and it is found that with inaccurate data, or with too few data points there is no justification for the extra parameter. However, even with inaccurate data, if sufficient points are measured it does become possible to estimate the damping constant, a useful lesson for the experimental laboratory.

6. The restricted three-body problem

As described above, in an earlier exercise the students are required to write their own programmes to solve the two-body problem (whose analytic solution is known). In this exercise a programme to solve the restricted three-body problem is provided.

In the restricted three-body problem the following assumptions are made:

(i) that two of the bodies are massive, the third of negligible mass;

(ii) that the two heavy bodies move in circular orbits about their centre of mass;

(iii) that the motion of all three bodies is coplanar.

Even with these restrictions the motion of the third body cannot be analytically calculated except in a few (five) special cases, known as the Lagrangian points.

The programme solves the equations of motion of the third body by the Runge-Kutta-Merson method which has a built in error check,[5] and the step length for integration is continually adjusted to keep the error term within preset limits. The current positions of the three bodies are displayed graphically: owing to the large size of the programme only a single colour display is used.

At each step of integration the programme calculates the osculating orbital elements of the third body. To do this it postulates that the third body is in a two-body (conic-section) orbit about the centre of mass, and from the instantaneous radius and velocity vectors calculates such quantities as the period and eccentricity of the conic section orbit with these vectors.

In general the osculating orbital parameters change slowly: in the two-body case they are constants. Usually they change slowly and periodically unless the third body approaches one of the other bodies (usually the second) closely, in which case its motion is seriously perturbed.

These orbital elements are continuously displayed, enabling the student to observe the changes.

6.1 *A transfer orbit*

The first exercise is to find analytically the transfer orbit for a satellite to be tangential to the (circular) orbits of the Earth and Jupiter, neglecting the gravity of the planets. Using the starting conditions of the analytic exercise, the motion of the satellite under the gravity of the Sun and Jupiter is observed, and the dramatic effect of changing the starting angle is studied. It is easy to eject the satellite out of the solar system by 'bouncing' it off Jupiter's gravitational field.

6.2 *The Lagrangian points* L_4 *and* L_5

In the rotating coordinate system, with origin at the centre of mass, there are five Lagrangian points: points at which a satellite will remain indefinitely under the gravitational forces of the two heavy bodies, and of these points are unstable in that a satellite tends to drift away from them. Of the remaining two points, L_4 and L_5, each lies at the vertex of one of the two equilateral triangles in the plane of rotation which have the two heavy bodies at the other two vertices. These points are stable; a satellite near one of these points remains near it indefinitely.

The motion of such a satellite cannot be determined analytically: its motion is almost epicycloidal, but the satellite does return repeatedly to the close neighbourhood of the Lagrangian point and never moves very far from it.

The nature of the motion of the satellite is best seen by plotting it in rotating coordinates, though even in this case the trajectory is liable to be very tangled at the end of a long run (a few hundred 'years'). The ability to replot the trajectory (an 'action replay') is valuable, since the student can then see the trajectory develop.

Summary

In the Computational Physics Laboratory the students carry out a variety of exercises illustrating some of the applications of computers in physics. In most of the exercises the availability of a graphics display is important: most of the exercises would be impossible, or impossibly tedious, without a graphics display.

Acknowledgements

I wish to thank the Department of Mathematics for allowing us to use their Computer Laboratory for this course. I also wish to thank numerous colleagues in both Physics and Mathematics with whom I have discussed these exercises, and particularly Rachel Howe for helping to implement the exercises.

REFERENCES

1 P.H. Borcherds: *Physics Education 1985* (to be published).
2 R.D. Harding: University of Cambridge Computing Practical Class Manual, Part 1B Mathematical Tripos (1982).
3 D.P. Landau and R. Alben: *American Journal of Physics* **41**, 394–400 (1973).
4 E. Whitaker and G. Robinson: *Calculus of Observations*. Blackie and Son (1924).
5 C.F. Gerald: *Applied Numerical Analysis*, 2nd edn. Addison-Wesley (1978).

Hybrid computer simulation of the thermal performance of a building

J. ROSELL AND R. MORGAN
Physics Department, University of Ulster, Coleraine, N. Ireland

Abstract: A network of resistors and capacitors is used to simulate a building. Weather is simulated by a microcomputer driving digital-to-analogue converters, which allow the generation of either ideal (sinusodial) diurnal variation of temperature and solar radiation or real hourly meteorological observations. The performance of the building is displayed by the microcomputer as graphs of temperatures and heat flows, and also as a total heating bill.

Introduction

There are many different techniques for simulation of thermal performance of a building, ranging from simple rule-of-thumb calculations to extremely complex computer models.[1,2] An electrical analogue has certain advantages, notably speed of operation, minimal computational requirements, and comprehensibility. Its main disadvantage is the difficulty of relating performance based on the regular waveforms produced by electronic function generators to the performance of the building when subjected to the much less regular waveforms produced by electronic function generators to the performance of the building when subjected to the much less regular behaviour of real weather. By using numerical data to generate the driving signal, it is possible to simulate real weather to quite good accuracy, simply by driving the model from a stored array of actual meteorological observations. The result is a hybrid digital/analogue computer model. The technique was developed by Day and Burberry[3] using paper tape readers and digital-to-analogue converters as the source of weather signals, and an analogue-to-digital converter connected to the analogue as the recording device. With the arrival of the microcomputer as a standard piece of laboratory equipment, this technique provides a very effective and economical alternative to purely digital techniques.

Equivalence between model and reality

In the simulation, temperature is represented by voltage, and heat flow by electrical current flow. Thermal resistance is represented by electrical resistance, and the thermal capacity of the building by capacitors. The quantitative relationship between model (electrical) and real (thermal) quantities depends on the relationship between real time and model time, the temperature–voltage equivalence, and the relationship between thermal and electrical resistance.

The following table summarises the equivalence relationship used in the present experiment.

Quantity	*Real unit*	*Present work*
Time	1 hour	0.5 seconds
	1 second	138.8 microsecond
Temperature	1 deg C	0.05 volts
Thermal resistance	1 deg C/W	1 megohm
Thermal conductance	1 W/deg C	1 microampere/volt
Quantity of heat	1 J	6.94 picocoulombs
Rate of heat flow	1 W	0.05 microamperes
Thermal capacity	1 J/deg C	138.8 picofarads

Circuit of the building model

According to Davies[5,6,7] it is possible to simulate a component of a building (such as a wall) to a high degree of precision by means of a relatively small number of electrical components. The array can be a seven-element R-C network for simulating the response to diurnal (sinusodial) temperature variations. Solar input is more closely approximated by a step change, and a five-element model is proposed for this situation. The concept has been extended by Basnett,[8] who proposes that a ladder network of seven components, four resistors and three capacitors, can achieve an extremely good fit at most of the reasonable likely exciting periods, which in Basnett's study were taken over a range from diurnal to 15 minutes. Thus it appears that resistance–capacitance models are capable of simulating buildings to quite satisfactory accuracy.

A block diagram of the system is shown in Fig. 1. The simplest model for the building consists of a resistor to simulate the thermal resistance of the building envelope, and a capacitor to represent the total thermal mass. Together with the connections to the driving and measuring circuitry, this is shown in Fig. 2.

A slightly more complex model is required for the most basic simulation of a cavity wall. In Fig. 3 there are capacitors representing the thermal capacity of the two brick walls and of the inside of the room, and they are connected by resistors R1 and R3 representing the thermal resistance of the walls (and also of the still air layer at the surface) and R2 representing the resistance of the cavity.

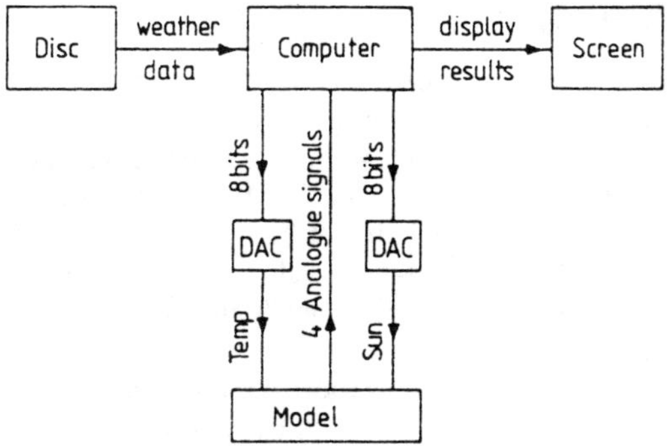

FIG. 1. Block diagram of the apparatus.

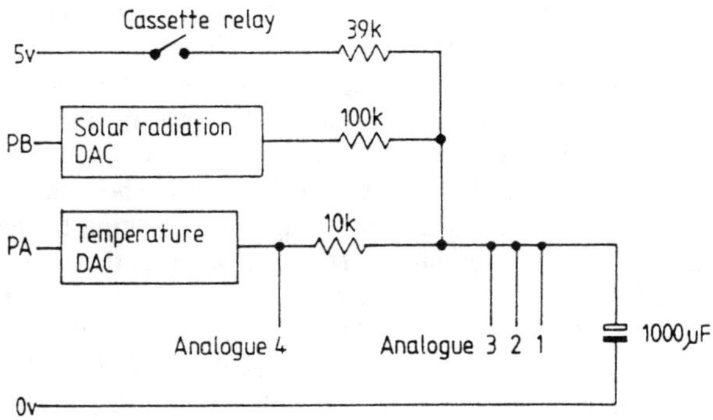

FIG. 2. Simple analogue model of a building.

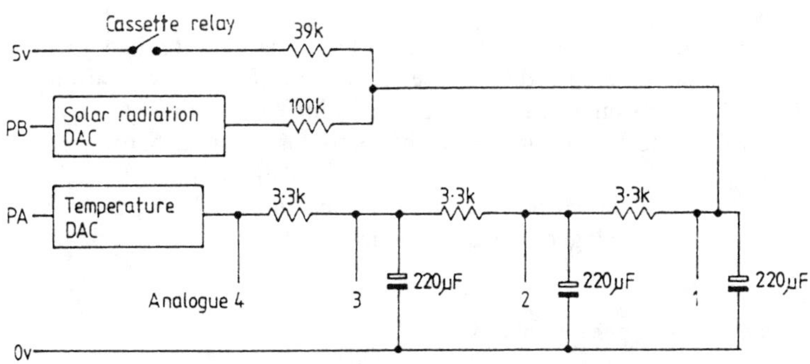

FIG. 3. Simple model of a cavity wall.

Measuring and driving circuit

Two climatic variables are generated by the computer, temperature and solar radiation, for which it is possible to use outputs from the two parallel ports of a BBC Microcomputer without additional engineering. Wind speed and direction are also relevant, but the extra electronics required adds considerably to the difficulty of setting up the experiment, and probably adds little to the teaching value. Further, an opportunity is created for the more able and electronically-minded student to devise the appropriate multiplexing circuit.

The output from the printer port (port A), representing temperature, is converted to an analogue voltage by an eight-bit digital-to-analogue converter. The full eight-bit range is used to minimise the step size, and the gain of the output amplifier is adjusted to make the full-scale output appropriate. A temperature range of 0°–25.5°C is a convenient choice, and a voltage range of 0–1.275 V provides an easy conversion relationship while staying within the 1.8 V of the analogue inputs.

Solar radiation is simulated rather crudely. The ideal representation is a current source whose magnitude is related to the solar radiation. This would require electronics of rather greater sophistication[3] than is justified in this experiment. Instead, a voltage source is used whose magnitude is sufficiently great compared with the voltages representing temperatures in the building and the environment (the sink voltage) that when passed through a high-value resistor, the current is approximately independent of the sink voltage. An eight-bit DAC driven by port B (the user port) provides the required signal. A radiation level of 600 W m^{-2} is represented by 240 and 4.8 V. When the voltage representing the solar radiation is low, the building can discharge into the DAC, simulating radiation from windows into a clear sky at night. If this is objectionable, a diode can be used to prevent it, equivalent to drawing the curtains at night.

As a simulation of a heating plant, the temperature (voltage) in the building can be compared with a set value, and if it is below, a relay is closed and turns on a current source consisting of a high resistance and a high voltage (5 V). To minimise the need for external hardware, the cassette relay of the BBC computer is used for the purpose. By keeping the voltages of the current sources to a maximum of 5 V, the risk of damage to the analogue inputs is minimised.

Selection of component values

Typical values for the resistors of Fig. 2 can be selected on the basis of the known thermal performance of buildings. A very well insulated small house maintained at an indoor temperature of 20°C with an outside temperature of −5°C might need a continuous heat input of about 2.5 kW. In practice the heating system is likely to be of much greater output than this, so as to allow for warming up in the morning after a cold night, but for the present purpose that can be ignored.

Thus a 25° temperature difference requires an input of 2.5 kW, which makes the thermal conductance 100 W/degree. The model equivalent for this is 10 kohms, which is a reasonable value.

Estimation of the thermal capacity is much more difficult, and the best procedure is to resort to experiment. By trying a range of capacitance values, reasonable results can be obtained from about 100 μF, representing a lightweight structure such as a timber-frame house, to 2200 μF, representing a heavyweight structure with thick concrete walls, roof and floor.

The previous estimate of a heat requirement of 2.5 kW under extremely cold conditions, is represented by 125 μA in terms of the electrical model. Making the assumption that we have about 5 V to drive the 'heating' current into the building, the resistance required is about 40 kohms. For the purpose of calculating a heating bill, we assume a price of 4 pence per kilowatt-hour, making the running cost 10 pence per hour.

The solar gain is a difficult quantity to estimate accurately. In full sun, a window may pass an average of about 200 W m^{-2} of window area, depending on the orientation to the sun's path. A typical small house might have 5 m^2 of window exposed to sun at any one time of day, giving a gain of 1 kW if the sun is at maximum intensity. In the electrical model we use 5 V to represent an approximation to full sun, and heat gain of 1 kW is represented by 0.05 mA. Thus the resistance required is approximately 100 kohms.

The estimation of the components of the ladder network of Fig. 3 is more difficult. The total resistance and capacitance can be calculated or estimated in the same way as for the simpler circuit of Fig. 2, but the distribution between separate elements is by no means obvious. Day and Burberry[3] give some examples which are said to be good representations of real thermal behaviour, and Davies[5] gives tabulated data.

Students can be invited to devise their own versions of the thermal circuit and try them out.

The program

A possible program, written in BBC Basic, is shown below. Although it makes use of some of the unique features of the BBC Microcomputer, it should be possible to use a similar approach for any computer equipped with a VIA or PIA for parallel input/out and an analogue-to-digital converter for anlogue voltage input.

```
 10 REM INITIALISE
 20 DEF FNS(T)=SIN(T*PI/12)
 30 DIM TEMP 170, SUN 170
 40 DIM OLFLOW(4),FLOW(4),IN(4),OLIN(4),FUJ(6)
 50 DIM SFLOW(2)
 60 SFLOW(1)=0:SFLOW(2)=0
 70 XSCALE=7:YSCALE=2:SUNSHIFT=700:BOILSHIFT=550:SAMPLE=5
 80 UP=&FE60:PP=&FE61:?&FE62=255:?&FE63=255
 90 SR=6:SS=18
100 FUJ(1)=YSCALE*250*1.8/(1.25*65520*SAMPLE)
110 FUJ(2)=YSCALE*250*1.8/(1.25*65520*SAMPLE)
120 FUJ(3)=YSCALE*250*1.8/(1.25*65520*SAMPLE)
130 FUJ(4)=YSCALE*250*1.8/(1.25*65520*SAMPLE)
140 FUJ(5)=1
150 FUJ(6)=1
160 FOR I=1 TO 4
170 OLFLOW(I)=0:FLOW(I)=0
180 NEXT I
190 REM MAIN PROGRAM START
200 MODE 7
210 PROCWEATHER
220 REM RE-CYCLE START
230 *MOTOR 0
240 BOILER=0 : REM HEATING OFF
250 ?PP=0:?UP=0 : REM START MODEL
260 MODE 7
270 PROCHEATING
280 MODE 4
290 @%=&00002
300 PROCTEMPGRAPH
310 PROCSUNGRAPH
320 PRINT TAB(2,11)"THERMOSTAT TEMPERATURE ";THERMSET
330 PRINT TAB(2,12) "PRESS G TO GO ON,"
340 PRINT TAB(8,13)"M TO GO BACK TO MENU"
350 G$=GET$
360 IF G$="M" GOTO 190
370 IF G$<>"G" THEN VDU7:GOTO 330
```

```
380 MODE 2
390 PROCDRIVE
400 PROCDUMP
410 MODE 7
420 PRINT TAB(0,10)"Another run with this weather (Y/N)?"
430 G$=GET$
440 IF G$="Y" GOTO 220
450 IF G$<>"N" THEN VDU7:GOTO 420
460 @%=&10
470 STOP
480 REM PROCEDURES
490 DEF PROCHEATING
500 PRINT TAB(12,2);CHR$(131);CHR$(141);"HEATING"
510 PRINT TAB(12,3);CHR$(131);CHR$(141);"HEATING"
520 PRINT TAB(0,7)"Set thermostat temperature"
530 PRINT TAB(0,9)"  (0 for no heating) . . . ";
540 INPUT THERMSET
550 TSET=THERMSET*YSCALE*10
560 ENDPROC
570 DEF PROCWEATHER
580 PRINT TAB(13,2);CHR$(131);CHR$(141);"WEATHER"
590 PRINT TAB(13,3);CHR$(131);CHR$(141);"WEATHER"
600 PRINT TAB(0,6)"To load ideal weather ............. I"
610 PRINT TAB(0,8)"To load ideal weather without sun .. J"
620 PRINT TAB(0,10)"To load real sunny weather......... S"
630 PRINT TAB(0,12)"To load real cloudy weather ........ C"
640 REPEAT
650 FL=0
660 G$=GET$
670 IF G$="I" THEN PROCFLAG:PROCLOADITEMP:PROCLOADISUN
680 IF G$="J" THEN PROCFLAG:PROCLOADITEMP:PROCLOADNOSUN
690 IF G$="S" THEN PROCFLAG:PROCLOAD("SUNNY")
700 IF G$="C" THEN PROCFLAG:PROCLOAD("CLOUDY")
710 IF FL=0 THEN VDU7
720 UNTIL FL=1
730 ENDPROC
740 DEF PROCFLAG
750 FL=1
760 PRINT TAB(14,15)CHR$(136);"LOADING.."
770 ENDPROC
780 DEF PROCLOADITEMP
```

```
790 FOR DAY=0 TO 6
800 FOR HR=0 TO 23
810 REM TEMP?(HR+DAY*24)=50*FNS(HR-8)+100
820 TEMP?(HR+DAY*24)=50*FNS(HR-8)+100+DAY*10
830 PRINT TAB(0,17)"TEMPERATURES .. DAY ";DAY
840 NEXT HR
850 NEXT DAY
860 TEMP?168=TEMP?167
870 TEMP?169=TEMP?168
880 ENDPROC
890 DEF PROCLOADISUN
900 FOR DAY=0 TO 6
910 FOR HR=0 TO 23
920 IF HR>SR AND HR<SS THEN SUN?(HR+DAY*24)=60*FNS(HR-SR)
ELSE SUN?(HR+DAY*24)=0
930 PRINT TAB(0,17)"SOLAR RADN .....DAY ";DAY
940 NEXT HR
950 NEXT DAY
960 SUN?168=SUN?167
970 SUN?169=SUN?168
980 ENDPROC
990 DEF PROCLOAD(M$)
1000 ONERROR CLOSE#Q:PRINT "DISC FAILURE":STOP
1010 Q=OPENIN(M$)
1020 FOR HR=0 TO 169
1030 INPUT#Q,TEMP?HR,SUN?HR
1040 NEXT
1050 CLOSE#Q
1060 ONERROR OFF
1070 ENDPROC
1080 DEF PROCLOADNOSUN
1090 FOR HR=0 TO 169
1100 SUN?HR=0
1110 NEXT
1120 ENDPROC
1130 DEF PROCTEMPGRAPH
1140 PROCSCALE
1150 MOVE 0,TEMP*YSCALE
1160 FOR HR=0 TO 168
1170 DRAW HR*XSCALE,TEMP?HR*YSCALE
1180 NEXT
```

```
1190 ENDPROC
1200 DEF PROCSUNGRAPH
1210 PROCSCALE
1220 MOVE 0,SUN*YSCALE+SUNSHIFT
1230 FOR HR=0 TO 168
1240 DRAW HR*XSCALE,SUN?HR*YSCALE+SUNSHIFT
1250 NEXT
1260 ENDPROC
1270 DEF PROCDRIVE
1280 FOR WK=0 TO 1
1290 CLS
1300 BILL=0
1310 PROCSCALE
1320 FOR HR=0 TO 168
1330 TIME=0
1340 ?PP=TEMP?HR : REM 25 C MAX
1350 ?UP=SUN?HR*4: REM 240 BITS=600W
1360 GCOL 0,1
1370 REM MOVE HR*XSCALE,TEMP?HR*YSCALE
1380 REM DRAW (HR+1)*XSCALE,TEMP?(HR+1)*YSCALE
1390 IF TSET<>0 THEN GCOL 0,1 ELSE GCOL 0,0
1400 MOVE HR*XSCALE,(BOILSHIFT+BOILER*50)
1410 DRAW (HR+1)*XSCALE,(BOILSHIFT+BOILER*50)
1420 GCOL 0,3
1430 MOVE HR*XSCALE,SUN?HR*YSCALE+SUNSHIFT
1440 DRAW (HR+1)*XSCALE,SUN?(HR+1)*YSCALE+SUNSHIFT
1450 FOR I=1 TO 4
1460 OLIN(I)=IN(I)
1470 IN(I)=0
1480 NEXT I
1490 FOR J=1 TO SAMPLE
1500 FOR I=1 TO 4
1510 IN(I)=IN(I)+ADVAL(I)
1520 NEXT I
1530 NEXT J
1540 FOR I=1 TO 4
1550 IN(I)=IN(I)*FUJ(I)
1560 NEXT I
1570 OLFLOW(1)=FLOW(1)
1580 OLFLOW(2)=FLOW(2)
1590 FLOW(1)=(IN(1)-IN(2))*FUJ(5)
```

```
1600 SFLOW(1)=SFLOW(1)+FLOW(1)
1610 FLOW(2)=(IN(3)-IN(4))*FUJ(6)
1620 SFLOW(2)=SFLOW(2)+FLOW(2)
1630 FOR I=1 TO 2
1640 GCOL 0,(I+4)
1650 MOVE HR*XSCALE,OLFLOW(I)+SUNSHIFT
1660 DRAW (HR+1)*XSCALE,FLOW(I)+SUNSHIFT
1670 NEXT I
1680 FOR I=1 TO 4
1690 GCOL 0,I
1700 MOVE HR*XSCALE,OLIN(I)
1710 DRAW (HR+1)*XSCALE,IN(I)
1720 NEXT I
1730 PROCCONTROL
1740 REPEAT UNTIL TIME>49
1750 NEXT HR
1760 NEXT WK
1770 @%=&020204
1780 PRINT TAB(3,12)"BILL=`";BILL
1790 REM PRINT TAB(3,3)"HEATIN";SFLOW(1)
1800 REM PRINT TAB(3,4)"HEATOUT";SFLOW(2)
1810 ENDPROC
1820 DEF PROCSCALE
1830 VDU5
1840 GCOL 0,7
1850 MOVE 0,0:DRAW 0,38*YSCALE
1860 FOR Y=0 TO 25 STEP 5
1870 MOVE 0,Y*YSCALE*10
1880 DRAW 20,Y*YSCALE*10
1890 PRINT Y
1900 NEXT
1910 PROCLABEL
1920 MOVE 0,SUNSHIFT
1930 DRAW 0,10*YSCALE+SUNSHIFT
1940 FOR Y=0 TO 10 STEP 5
1950 MOVE 0,Y*YSCALE*10+SUNSHIFT
1960 DRAW 20,Y*YSCALE*10+SUNSHIFT
1970 PRINT Y
1980 NEXT
1990 VDU4
2000 ENDPROC
```

```
2010 DEF PROCDUMP
2020 G=GET :REM INSERT DUMP PROGRAM HERE
2030 ENDPROC
2040 DEF PROCCONTROL
2050 IF TSET<>0 AND IN(1)<TSET THEN BILL=BILL+0.1:BOILER=1
ELSE BOILER=0
2060 IF BOILER=1 THEN *MOTOR 1
2070 IF BOILER=0 THEN *MOTOR 0
2080 ENDPROC
2090 DEF PROCLABEL
2100 IF G$<>"G" THEN ENDPROC
2110 MOVE 0,585:PRINT "OFF"
2120 MOVE 0,635:PRINT "ON"
2130 MOVE 150,50:GCOL 0,1:PRINT "ROOM T"
2140 MOVE 700,50:GCOL 0,4:PRINT "OUTSIDE T"
2150 MOVE 150,950:GCOL 0,5:PRINT "INNER"
2160 MOVE 550,950:GCOL 0,6:PRINT "OUTER"
2170 MOVE 950,950:GCOL 0,3:PRINT "SUN"
2180 GCOL 0,7
2190 ENDPROC
```

This program can be used without 'real' weather if the 'ideal' options are selected from the menu. If the options to use 'real' weather are required, it will be necessary to have data files called 'SUNNY' and 'CLOUDY'. The required format is a sequence of 170 pairs of integers, each between 0 and 255. The first number of each pair is the Celsius temperature in tenths of a degree (250=25°C), and the second is the solar radiation per square metre (240=600W).

Acknowledgements

The authors acknowledge the support of the Garfield Weston Trust, the McCrea Research Fund and the Science and Engineering Research Council.

REFERENCES

1 J.K. Whittle: *A study of consumption in housing using computer stimulation*. PhD thesis, University of Newcastle on Tyne (1979).
2 P.W. O'Callaghan:*Building for Energy Conservation*, p. 232. Pergamon Press (1978).

3 B.F. Day and P.J. Burberry: The application of analogue techniques to the study of the thermal performance of buildings, *Proc. First International Symposium on Energy Conservation in the Built Environment; Conseil International du Batiment* (R.G. Courtney Ed.), pp. 282–305. Construction Press, Lancaster (1976).

4 F.M. Wai, K.M. Letherman and P.J. Burberry: Validation and calibration techniques for hybrid computer simulations in the prediction of seasonal energy consumptions, *Proc. Third International Symposium on Energy Conservation in the Built Environment; Conseil International du Batiment* (P. Minogue Ed.), pp. 2.1–2.10. An Foras Forbartha, Dublin (1982).

5 M.G. Davies: Optimal RC networks for walls, *Appl. Math. Modelling* **6**, 403–404 (1982).

6 M.G. Davies: Optimum design of resistance and capacitance elements in modelling a sinusoidally excited building wall, *Building and Environment* **18**, 19–37 (1983).

7 M.G. Davies: A design procedure to model a transiently excited wall, *Appl. Math. Modelling* **7**, 182–188 (1983).

8 P. Basnett: A new model for calculating building energy requirements, *Proc. Symposium on the Environment inside Buildings*, Inst. of Mathematics and its Applications, Electricity Council Research Centre, Capenhurst, February, 1985 (to be published).

Shoestring closed-loop computer control in FORTH

D.G. SMITH

Department of Physics, University College of North Wales, Bangor LL57 2UW

Abstract: A simple low-cost apparatus for an undergraduate laboratory is described. A conveyor track fitted with reflective sensors moves bar-coded blocks under computer control. The apparatus serves to introduce the basic ideas of computer control, and the language FORTH. Alternative operation in BASIC allows a valuable comparison of languages to be made by the student.

Computing activity in the physics degree scheme consumes time and resources. It is important that the integration of computing into the study and practice of physics be efficient in terms of both time and cost. Opinions will vary as to how this should be done, but general agreement could probably be reached on the importance of the control loop. This is fundamental to the control of instruments, machines and processes which may be represented as shown in Fig. 1. There would probably be less agreement on the question of which programming language to use for the control programs. To take two extreme cases, the use of BASIC often leads to programs that are lengthy, poorly structured and slow in execution, whereas the use of ASSEMBLER is tedious and time-comsuming for all but the simplest of tasks. Program maintenance is also a problem in both languages.

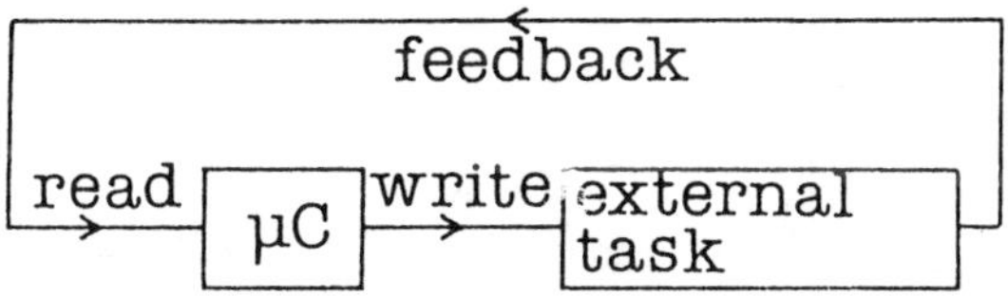

FIG. 1. The control loop.

The apparatus described here has two principal functions. Firstly, it introduces FORTH, a language that allows fast compact control programs to be written and maintained efficiently. Second, it introduces and illustrates the control loop in one of its simplest mechanical forms, namely the conveyor track. The use of a mechanical system is important. It is my experience that, other factors being equal, more learning 'mileage' is available from apparatus that moves than from apparatus that does not. The physical layout is shown schematically in Fig. 2. A reversible d.c. motor controlled by a microcomputer is coupled through a gearbox to the conveyor track. The track carries aluminium blocks, each provided with a simple bar-code in the form of four 10 mm wide vertical bars of either matt-black point or polished metal. The bar-code of a given block can be sensed by any one of eight infra-red reflective opto-switches sited along the track as shown in Fig. 3. The signals from the opto-switches can be read by the microcomputer.

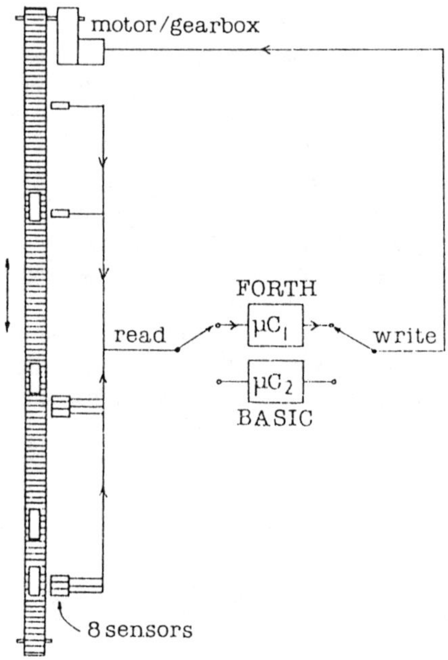

FIG. 2. The conveyor track.

Though the system is conventional, there are one or two points of design that have bearing on the question of operational reliability. This is important because of the inconvenience and frustration that equipment failure or malfunction during a laboratory session can cause. Microcomputers and their peripherals are generally trouble-free provided that they are not being frequently connected, disconnected and moved

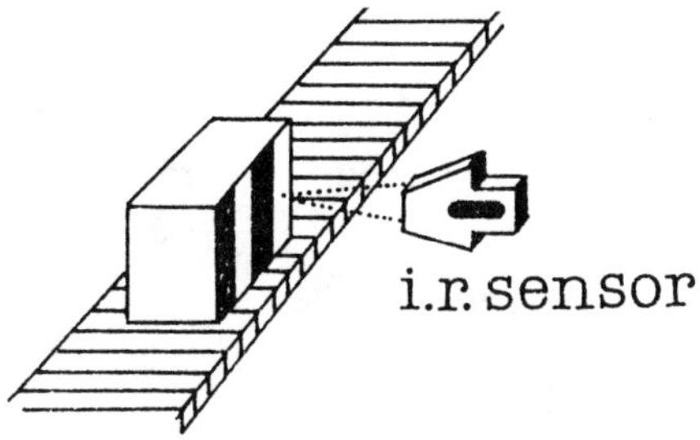

FIG. 3. Bar-code sensing by means of reflective opto-switches.

around. In this important respect it is preferable to use a very cheap microcomputer for laboratory work, because it can then become a permanent part of the apparatus. As a bonus, the low-priced microcomputer will also probably be relatively small physically, and therefore will be less likely to be the dominant component of the apparatus. The conveyor track must move smoothly and without lateral movement on its supporting surface. Most flexible flat-belt arrangements will probably be found to be unsatisfactory. Our solution was to use LEGO plastic-link tracking. This is inexpensive and has worked perfectly without maintenance or adjustment for more than two years. The tracking has a total length of about 1.2 m, and slides over an aluminium supporting surface.

The control system is shown in block form in Fig. 2. Two cheap microcomputers are included. For FORTH the Jupiter ACE is used. This machine is ideal for student use. The FORTH is ROM-based and is therefore quite student-proof. For BASIC, a Sinclair ZX-81 has been used, although a Sinclair SPECTRUM would probably have been a better choice. The 8-bit (TTL) IN and OUT lines of each microcomputer may be connected to the conveyor track electronics in turn by means of a 16-pole wafer switch. The WRITE electronics for controlling the electric motor is shown in Fig. 4. The track may be operated at one of three speeds, 12 mm s^{-1}, 18 mm s^{-1} or 30 mm s^{-1}, in each direction. A bit-controlled flashing LED is also provided. The READ electronics for the eight reflective opto-switches is shown in Fig. 5. One worthwhile though non-essential feature incorporated is the display of the state (0 or 1) of a given opto-switch by means of a small LED mounted on the opto-switch rather than on the circuit board. This allows instant confirmation of the READ information presented to the microcomputer. Unreliable machine reading can then be identified immediately, removing any uncertainty on the student's part as to whether a problem is hardware-based or software-based.

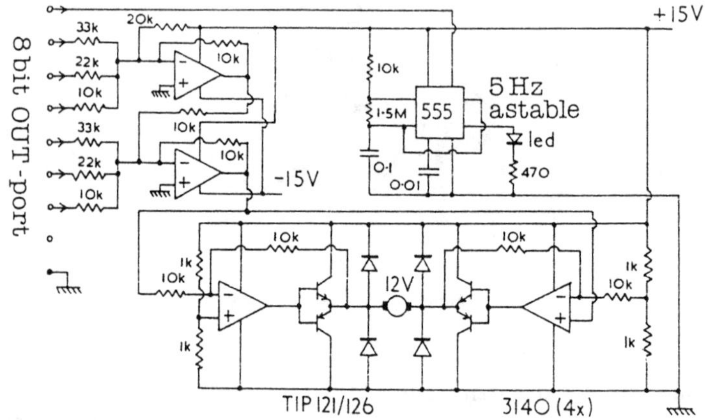

FIG. 4. The WRITE electronics.

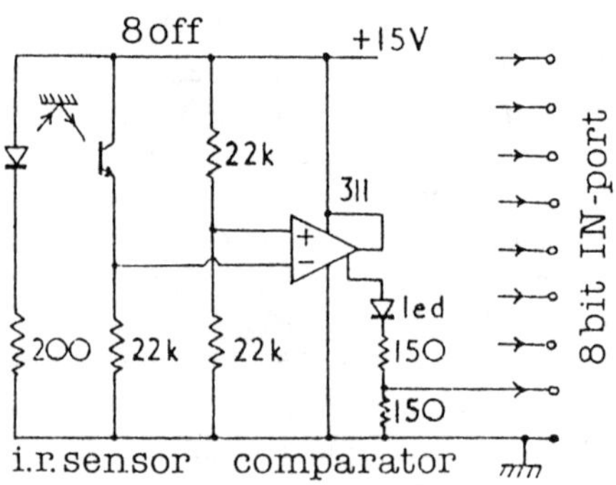

FIG. 5. The READ electronics.

The total cost of the apparatus, including the two microcomputers and their interfaces is approximately £220 at 1985 prices. A UHF monitor and a ±15 V, 1 A d.c. power supply will also be needed. Exactly how the apparatus is used by the student should, as always, reflect the views and priorities of the laboratory class supervisor. Programming a computer for real-time external tasks introduces valuable aspects of computing that are not present in mathematical or data-processing tasks. If the student has not used FORTH before, then she will need two or three afternoon's work with an ACE to penetrate the language. The manual for the ACE is very good, and will serve both as a FORTH tutor and as a general reference source for the machine. It is my experience that physics students have very little difficulty with FORTH, and are often spontaneously complimentary about the language.

A wide range of graded exercises in computer control is possible with this simple apparatus. The program execution is relatively fast in FORTH. It is possible to generate closed-loop oscillations at about 10 Hz. Even so, the apparatus does not do justice to the speed of FORTH, and a separate control topic in which this speed is fully exploited would complement the conveyor track. Such a system could also make use of the software facilities of one of the several multi-tasking FORTHs now available for microcomputers.

Note added in proof

Soon after the 1985 Exhibition, the original two microcomputers of the apparatus were replaced by a single Amstrad CPC 6128 CP/M machine. This has a built-in disc drive and the conveyor track can now be operated from the one keyboard in either FORTH or BASIC (or PASCAL), as desired. The same facilities could of course be made available, at some extra cost at present, using any standard PCDOS/MSDOS system.

An introduction to on-line computing

J. Lowell and M.D. Lumb
University of Manchester Institute of Science and Technology, Manchester M60 1QD

Abstract: This 2-day project provides a bridge betweeen 'computational' and 'control' applications of microcomputers, and provides a simple introduction to such matters as ports, A-to-D conversion, bit masking etc. to students who have already acquired some knowledge of BASIC.

1. Introduction

We describe a simple project which provides an introduction to on-line computing for 1st year students in preparation for more advanced work later in the course. We are keen to show them that even an elementary knowledge of a simple language (BASIC) is sufficient to achieve quite sophisticated objectives in data logging and control. The project is entirely concerned with software; the necessary electronic interfacing is provided (we explain that the electronics knowledge necessary to construct such interfaces will be acquired in the second year of the course).

Students write a program to control a 'phototropic robot'. The details will be clear from the instruction sheet which we reproduce (in part) below. A common difficulty with this kind of teaching arises from the great variation in experience among students. Some are almost novices while others are experienced hobbyists. We set up a simple task (to control the system by fairly crude feedback) which all students can manage; however, the more experienced can go on to write a program to implement a three-term proportional controller.

The instruction sheet which follows is an edited version of the one issued to our students — we have omitted some 'hints' about the subroutines implementation as well as a preliminary section describing port addressing, bit masking and so forth.

2. Instructions

First note carefully that the practical exercise you will do in the laboratory is designed to teach you *computing*, not electronics. All the necessary equipment will be provided ready-made and you should *not* concern yourself about how it works. All you need to know is the information necessary to write the computer program.

Your task will be to program a light-seeking robot, such as might be used to make solar-power equipment 'track' the Sun across the sky.

The apparatus shown in Fig. 1 consists of two electronic 'eyes' mounted on the shaft of a d.c. motor; the computer monitors the intensity of light at each eye and if it finds one eye is receiving more light than the other it turns the motor in the appropriate direction, until both eyes are equally illuminated. In all servosystems of this kind it is desirable to control the speed of the motor rather than turning it off or on. Therefore, a D-to-A converter is used to generate different speeds. The 'eyes' of the robot produce a voltage proportional to the intensity of the light they 'see'. After suitable amplification this voltage can be fed directly to an A-to-D converter connected to the computer. However, there are *two* eyes; we could, in principle, use a separate A-to-D and port for each eye. But this is an expensive and clumsy method; it is much better to use a device called a *multiplexer* to switch between the eyes. This multiplexer is essentially a two-pole switch whose position is controlled by a 'selector' voltage (the position changes if the selector voltage changes from 0 to 5 V or vice-versa). The equipment described so far is *in principle* sufficient to do the job. However, it would be easy to make programming errors resulting in the motor turning round and round continually, tangling and breaking the leads to the 'eyes'. To avoid this problem (and for convenience in other ways) a 'limit detector' is provided; this is a simple switch connected to one bit of port B. If the motor is at the limit of its turning (in either

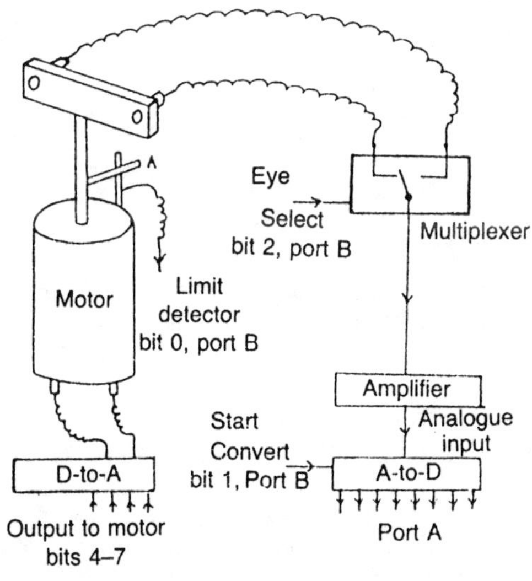

FIG. 1. Apparatus for light-seeking robot.

direction) this bit is connected to 0 V. Otherwise (if the arm A is *not* touching the limit detector) it is connected to 5 V.

The following information (refer to Fig. 1 also) is all you will need to program the equipment:

1. *Multiplexer*

If Bit 2 of port B is a '1' the left eye is selected. If it is a '0' the right eye is selected.

2. *Analogue to Digital (A-to-D) converter*

This device has to be told to convert its analogue input to digital form. It therefore has a second input voltage called 'Start Convert'(S.C.).

To initiate a conversion this voltage must be changed from 5 V to 0 V and *back again*. The voltage at the analogue input is then converted to binary form and appears at the output. This digital number at the output will not change until another SC pulse is received. Therefore, in order to convert the eye-voltage to binary form you must

(a) set bit 1 of port B to a '1' (unless it is a one alredy)
(b) set it to 0
(c) set it to 1 again

The value of the eye-voltage can then be read from port A, by the usual read instruction.

3. *Limit detector*

This is simple. If bit 0 of port B is a 0, the motor is at its limit. Otherwise it is not.

4. *Digital to Analogue (D-to-A) converter*

The output of this device is a voltage V whose size depends on the digital number N supplied to its inputs in accordance with the formula

$$V = N - 128$$

Note that only bits 4–7 are available for N (1–3 are used for other purposes). Thus N can take the values 16, 32, 48 ..... up to 240. It should be clear that:

(i) $N = 128$ means the motor is switched off.
(ii) $N = 112$ gives the slowest speed in one direction and $N = 144$ gives the slowest speed in the opposite direction.

(iii) N = 96, 160 are the next-fastest speeds in the two directions.
(iv) N = 16 and N = 240 are the fastest possible speeds in the two directions.

(Negative values of V mean *clockwise* turning of the motor.)

5. *Ports*

Port A is used exclusively to read the eye-voltage and it should therefore be programmed to be an INPUT port. All bits of port B except bit 0 are used for output, but bit 0 is used for input. Port B should be programmed accordingly.

6. *The Sun*

Since the experiment is to be performed in Manchester the Sun will be simulated by a light bulb.

Writing a programme

The VITAL technique in writing a control program is to *compartmentalise* it, that is, to write a separate program (a subroutine) for each task (such as, read the voltage from each eye). Each facet of the task can thus be tackled separately and independently of the others, and the basis of a working program gradually built up. This 'structured' approach is beneficial in any kind of programming but it is especially important in a control program since interaction with the outside world makes debugging more difficult, and an unforeseen error may in general cause physical damage to the equipment. However, do not worry too much about damaging the apparatus provided as we hope it is 'student proof'.

You should begin, therefore, by writing subroutines to achieve precisely defined (and simple objectives). It will then be a simple matter to construct a program by 'calling' these subroutines as required. It will be convenient to being the first subroutine at line 1000, the next at 2000 and so on. This leaves statements 0–999 for your actual program.

The subroutines you will need are:

(1) SWITCH ON MOTOR AT REQUIRED SPEED
(2) MOVE TO REFERENCE POSITION (sub. 2000)
(3) READ BOTH EYES (sub. 3000) LE, RE
(4) SCAN AND FIND MAX
(5) TRACK THE SUN

Further work (time permitting)

(i) The system is likely to 'lose' the Sun — e.g. when it is obscured by a cloud — and when this happens its behaviour will be unpredictable. Modify your 'track' routine to take this possibility into account.
(ii) (more difficult) The accuracy of the system can be enhanced by including 'integration' in the control loop. In the 'track-the-Sun' routine above (LE-RE) may be non-zero without the motor turning on (if the gain is small) causing error in direction.
(iii) (even harder) If you increase the gain too much in an attempt to make a quick and accurate controller the system is likely to oscillate. You can avoid this by introducing damping — either by 'hardware' (a bucket of oil) or by 'software'.

Building a Z80 minimum system computer

J.C. Matthews
Department of Physics, University of Nottingham, Nottingham NG7 2RD

Abstract: In order to gain a good basic understanding of microprocessor hardware, students wire up a simple Z80 minimum system computer using only 4 chips. After being carefully guided through the initial construction they go on to extend the computer to their own requirements using a range of other 'bits and pieces'.

Introduction

It may not be 'real physics' but it is a fact that today's physics graduates are expected to have a good understanding of the operation of microcomputers and microprocessors. Of course most of them will have used computers for number crunching, graph plotting, word processing etc., and some will have interfaced other quipment to them perhaps via a 'user port' (if such a port is in fact present on the computer in question) or even via the 'system bus'. However, to the vast majority of physicists, the sight of an immaculate printed circuit board under the lid of a personal computer is awe inspiring and usually produces a conviction that the hardware is extremely complex and probably never to be understood. The printed circuit board itself gives the impression that it is an integral part of any computer and that a normally wired circuit, of the 'birds nest' variety, could never work satisfactorily.

This may be compared with the situation facing a budding amateur bricklayer who decides to build a small barbecue. He first looks at a multistorey office block and quickly decides that building is all far too complex for him even to make a start.

The purpose of this experiment is to reduce the apparent complexity of microprocessors to its lowest possible level, to introduce the student as gently as possible to computer hardware and to show him that it is very easy indeed to build a simple working computer using 'birds nest' wiring. It then becomes a simple matter (a) to program it, and (b) to extend the hardware.

Of course the student does not make up a complete desk-top personal computer with keyboard, screen display, floppy disk interface etc. In fact, in order to help him to understand the fundamental operation of the microprocessor he need only make up the simplest possible computer, a 'minimum system'. This consists of four inexpensive chips plus a single capacitor, and yet, once it is wired up and checked, it forms a genuine computer capable of being programmed to carry out a vast variety of tasks.

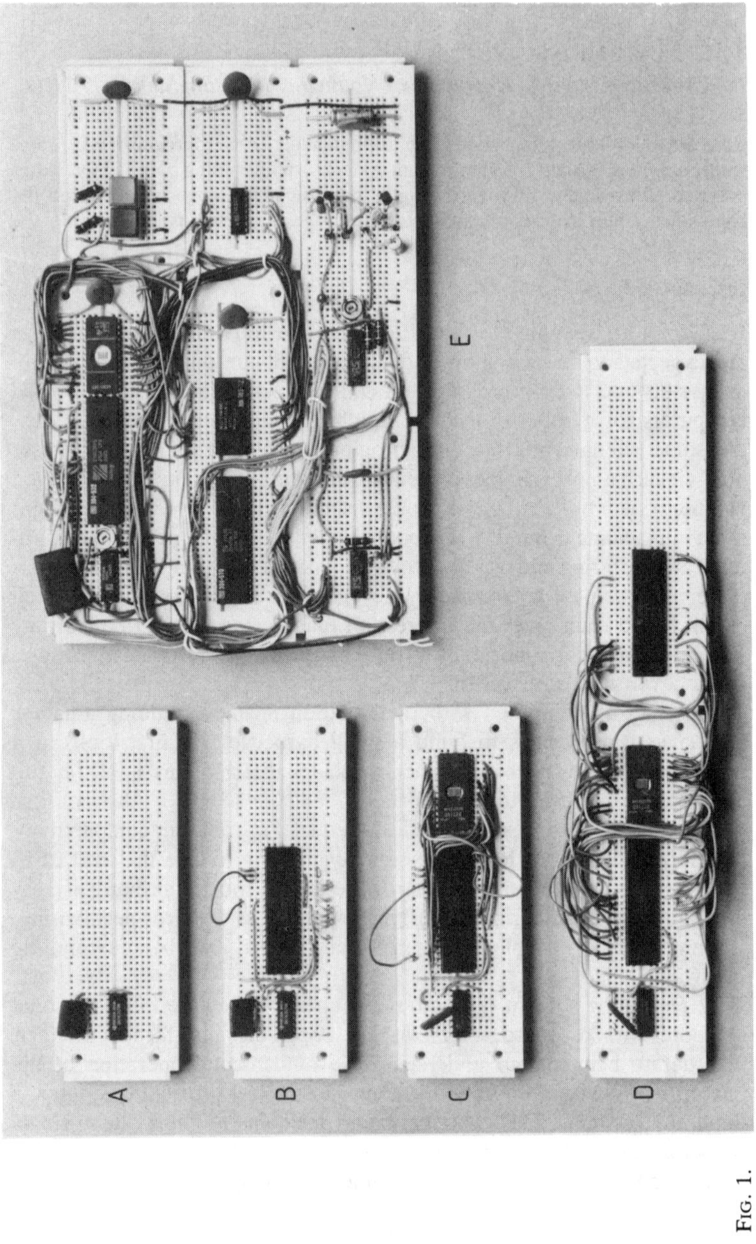

FIG. 1.

The student is carefully guided through this part of the experiment where he makes up the computer in four stages. At each stage he is told to examine particular signals in his developing circuit so that by the time the computer is completed he is fully capable of fault finding, if this becomes necessary.

Organisation of the experiment

The Z80 microprocessor was chosen on the grounds of economy and familiarity although the 6502 would have been equally suitable.

Each pair of students electing to do this experiment is given two bread boards, a set of four basic chips plus a capacitor, a power supply and an oscilloscope along with detailed instructions on the construction of the minimum system.

The four chips are:

(a) a 74625 voltage controlled oscillator, the 'clock'.
(b) a Z80 CPU microprocessor.
(c) a 2716 EPROM in which two test programs have been written.
(d) a Z80 PIO, parallel input–output device.

In addition to these common components, a variety of additional devices (keyboards, A-D & D-A converters, serial chips etc) are available for use in the second part of the project.

Construction of the computer

The computer is built up in four distinct stages:

Fig. 1A. *Building the clock*

The 74625 VCO requires only four connections to ground, three connections to 5 V, and the addition of a small capacitor to make up a clock with a frequency in the range 0.1–2 MHz (depending upon the value of the capacitor). It takes about 2–3 minutes to complete this stage of the construction and to measure the frequency of the clock.

Fig. 1B. *Connecting the Z80 CPU and running a NOP program*

16 connections have to be made in this stage taking about 15–25 minutes. No other components are required.

The 8 data lines of the Z80 are all connected to ground at this point so that the microprocessor thinks it sees a memory device filled with NOPs, op-code 00.

In order to start the 'NOP' program the Z80 is reset by simply touching the RESET lead temporarily to ground and then resetting it to 5 V. The aim here is to show the essential simplicity of the device and that it is not even necessary to use a switch to perform the reset.

By looking at the clock the M1 line of the Z80-CPU on the oscilloscope it is easy to see that there are four T-states in each NOP instruction.

It is also easy to see that the address bus is simple incrementing (by looking in turn at address lines A1, A2 etc) as the Z80 blindly continues to fetch NOPs from the nebulous memory.

Fig. 1C. *Connecting the EPROM*

There are 23 connections to be made in this stage taking about 20–30 minutes.

The student is given an EPROM in which two programs have been written. The first of these is written at address 0000 and is simply a jump to 0000. This program is always executed when the Z80 is at reset and it can be observed by looking at the clock and the M1 lines where it is seen that there are 10 T-states in each JMP instruction.

The purpose of the second program is to demonstrate the action of an interrupt or rather an NMI (non-maskable interrupt). It consists of a loop of five instructions written at address 0066, which is the address called when an NMI is performed. Again, to show the essential simplicity of the NMI interrupts on the Z80, it is only necessary to temporarily touch the NMI lead to ground. No switch or other circuitry is needed.

By looking at the clock and M1 lines it is quite easy to see the repeated pattern of five instructions in the loop.

The number of T-states in each instruction in the loop can be easily checked so that the student can see immediately the relationship between the number of T-states in a sequence of instructions and the speed of execution of these instructions.

Other signals that may usefully be examined at this stage are the MREQ, IORQ, WR and RD lines of the Z80, and these may be correlated with the instructions that are being executed.

Fig. 1D. *Connecting the Z80 PIO*

Only 18 connections have to be made taking about 20–30 minutes.

The Z80-PIO is a two-port parallel input/output chip and the second test program mentioned above is also used as a test program for this device.It simply outputs a square wave on each pin of port A of the PIO so that an immediate check can be made to see that it has been correctly connected.

Programming a new EPROM

Having completed the minimum system it is now a useful exercise to reprogram the EPROM. Typical programs might use port B as an input port to check one or more logic signals while port A can be made to respond to these different inputs. For example, an inexpensive crystal earphone may be connected directly to any output pin of the PIO and the EPROM may then be programmed to produce a set of tunes selectable via the other (input) port of the PIO. Alternatively, a set of different pulse sequences may be output, depending on the input to the PIO. Although these applications are trivial in themselves they emphasise the essential simplicity of the system and the ease with which it can be constructed and reprogrammed.

Fig. 1E. *Extending the minimum system*

The first addition to the minimum system is usually a 1 Kbyte-wide static RAM. By adding a RAM it becomes possible to make use of subroutines in a program since it is now possible to hold a stack. The simplest possible (ambiguous) address decoding is used making use of a single 7400 to provide the necessary signals to distinguish between the CE pins of the two memories.

Each pair of students then go their own way in further developing the system and it is usually found that they are perfectly capable of adding extra devices such as A-D and D-A converters, LEDs and 7-segment displays, keyboards etc. Figure 1E shows a minimum system to which has been added a static RAM, an analogue to digital (A-D) converter and a digital to analogue converter (D-A), to make up a useful transient recorder.

Examples of systems developed in this way include:

1. a noughts and crosses machine.
2. a pair of stepper motor drivers to control a small vehicle.
3. a transient recorder/signal averager (Fig. 1E).
4. a rotating message display.
5. a multistorey lift control.
6. a bar code reder.
7. a change giving vending machine.
8. a level crossing control.
9. intruder/fire controller.
10. single board computer with hex keyboard/7-segment display.

Conclusions

Students are pushed in at the deep end and asked to construct a minimum system computer, usually before they have attended any lecture course on the subject. As a result it has been found that their motivation to learn both assembly and language and interfacing has been greatly enhanced. Their timidity concerning computer hardware has been very much removed and they are far more comfident about tackling interfacing of real computers. They have a greater understanding of the operation of both the Z80 and also the other microprocessors (such as the 6502) along with other related devices and are able to write good efficient assembly language programs.

Electronics is generally regarded as being an essential tool for all physicists. Although it is not 'real physics', microelectronics is an important extension of this tool and the experiment described above is a quick and easy way of learning the fundamentals of both the hardware and the software aspects of this increasingly important topic.

Two simple VELA experiments

S.J. Parkinson,
Ulverston Victoria High School, Springfield Road, Ulverston, Cumbria LA12 0EB

Abstract: The VELA is a microprocessor-based instrument which can be used to measure and record potential difference, time and frequency. It enables many standard 'O' and 'A'-level experiments to be performed more easily and effectively. The two experiments are examples of this.

Sonometer experiments

The standard school experiments to investigate the variation of frequency of vibration of a wire with variation in length and tension of the wire can be performed using the frequency measuring programme 04 on the VELA.

A sonometer wire has a small piece of tape fastened to its centre. On either side of the tape are located a light source and a photodiode. They are arranged at the same vertical height as the tape, so that the light just shines across the tape onto the photodiode. When the wire is vibrated vertically, the light beam to the photodiode is cut at the same frequency as the vibrating wire. A varying potential difference is thus produced at the photodiode which has a frequency equal to that of the vibrating wire. Using programme 04, the frequency of this varying potential difference can be read and displayed by VELA. The frequency appears on the VELA display and is updated every second, hence the readings start at a low value, build up to a maximum value, which is constant for a few seconds, and then decreases as the vibration of the wire dies away.

There is an upper limit of about 200 Hz to the frequency which can be accurately read. This is not a limitation of VELA, which can read up to 20,000 Hz, but because the amplitude of vibration of the wire above a frequency of 200 Hz is not sufficient to cut the light beam.

SHM of an oscillating mass on a spring

The experiment examines the position of an oscillating mass on a spring as a function of time. A light is shone through the spring, near to its point of suspension, onto a photo-diode. The intensity of the light incident on the photo-diode depends on the extension of the spring. The p.d. across the photo-diode depends on this intensity and is measured using VELA.

The first part of the experiment measures the static value of the p.d.

Results

TABLE 1.

Tension in N	$\sqrt{T}$ in $N^{\frac{1}{2}}$	f in Hz
55	7.42	92
45	6.71	85
35	5.92	72
25	5.0	68
15	3.87	53
5	2.24	31

Error in f of ±2 Hz.

TABLE 2.

Length in cm	$1/l$ in cm^{-1}	f in Hz
100	0.0100	74
90	0.0111	85
80	0.0125	98
70	0.0143	107
60	0.0167	122
50	0.0200	144
40	0.0250	173

against the extension of the spring using the VELA programme 00, the digital voltmeter programme.

The results indicate a non-linear relationship over the full range of illumination. However, when the spring is oscillated, the amplitude can be kept sufficiently small to keep the intensity readings on the first part of the graph, which is linear within the errors of the measurements.

The second part of the experiment measures and records the p.d. every 20 ms as the mass oscillates using the VELa programme 02. The p.d. is then converted to position using the static measurements and a graph of position against time plotted. The p.d measurements may also be displayed on a CRO as usual.

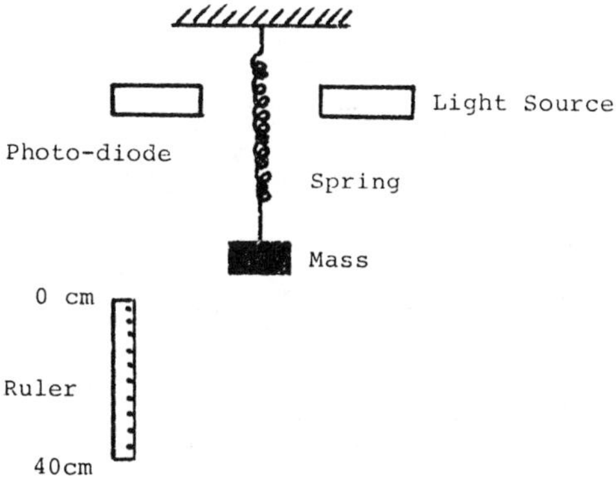

FIG. 1.

Comments

(a) Only every fifth value of position has been plotted.
(b) The static values of p.d showed quite large variation estimated as ±0.10 V.
(c) The experiment was initially performed using the light source powered from an a.c. supply. This gave a ripple on the p.d. measurements of a significant size. The light source was subsequently powered from a d.c. supply.
(d) As a rapid demonstration of SHM the CRO display is an adequate indication.
(e) The photo-diode was powered using three 1.5 V dry cells. Using the VELA in position (1) the readings are between 0 and 4.5 V, hence the ±25 V range needs to be used. By connecting the black VELA input as shown in (2), the ±2.5 V range can be used.

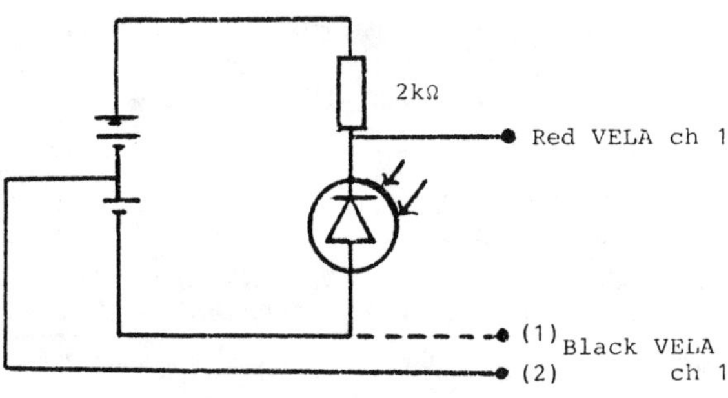

FIG. 2.

Results

Table 3. Static value of p.d. against position.

P.d. in V	2.7±0.10	2.50	2.10	1.75	1.45	1.27	1.10	0.82
Position in cm	0.0±0.5	2.8	5.5	8.2	11.0	13.7	16.4	19.0

P.d. in V	0.82	0.50	0.35	0.25	0.23	0.14	0.10
Position in cm	21.9	24.5	27.3	29.9	32.7	35.0	38.0

Table 4. P.d. and position against time.

P.d. in V	2.16	2.30	2.44	2.46	2.42	2.48	2.48	2.44	2.42	2.48
Position in cm	5.4	4.0	2.6	2.4	2.8	2.2	2.2	2.6	2.8	2.2
Time in ms	20	40	60	80	100	120	140	160	180	200
P.d. in V	2.46	2.42	2.30	2.16	1.93	1.76	1.64	1.48	1.40	1.38
Position in cm	2.4	2.8	4.0	5.4	7.7	9.4	10.6	12.2	13.0	13.5
Time in ms	220	240	260	280	300	320	340	360	380	400
P.d. in V	1.23	1.25	1.09	1.21	1.25	1.25	1.23	1.33	1.33	1.33
Position in cm	14.7	14.5	16.1	14.9	14.5	14.5	14.7	13.7	13.2	13.7
Time in ms	420	440	460	480	500	520	540	560	580	600
P.d. in V	1.27	1.42	1.42	1.56	1.62	1.66	1.79	1.93	2.06	2.07
Position in cm	14.3	12.8	12.8	11.4	10.8	10.4	9.1	7.7	6.4	6.3
Time in ms	620	640	660	680	700	720	740	760	780	800
P.d. in V	2.16	2.28	2.30	2.30	2.38	2.40	2.42	2.44	2.40	2.34
Position in cm	5.4	4.2	4.0	4.0	3.2	3.0	2.8	2.6	3.2	3.6
Time in ms	820	840	860	880	900	920	940	960	980	1000
P.d. in V	2.32	2.28	2.18	2.16	2.05	1.93	1.79	1.66	1.62	1.56
Position in cm	3.8	4.2	5.2	5.4	6.5	7.7	9.1	10.4	10.8	11.4
Time in ms	1020	1040	1060	1080	1100	1120	1140	1160	1180	1200
P.d. in V	1.50	1.56	1.46	1.37	1.40	1.38	1.27	1.21	1.19	1.17
Position in cm	12.0	11.4	12.4	13.3	13.0	13.2	14.3	14.9	15.1	15.3
Time in ms	1220	1240	1260	1280	1300	1320	1340	1360	1380	1400
P.d. in V	1.19	1.35	1.33	1.42	1.46	1.50	1.65	1.68	1.76	2.03
Position in cm	15.1	13.5	13.7	12.8	12.7	12.0	10.6	10.2	9.4	6.7
Time in ms	1420	1440	1460	1480	1500	1520	1540	1560	1580	1600
P.d. in V	2.18	2.26	2.28	2.42	2.48	2.48	2.48	2.48	2.44	2.32
Position in cm	5.2	4.4	4.2	2.8	2.2	2.2	2.2	2.2	2.6	3.8
Time in ms	1620	1640	1660	1680	1700	1720	1740	1760	1780	1800
P.d. in V	2.32	2.26	2.07	1.99	1.83	1.66	1.50			
Position in cm	3.8	4.4	6.3	7.1	7.7	8.7	10.4	12.0		
Time in ms	1820	1840	1860	1880	1900	1920	1940	1960		

The error in p.d. is taken as ± 0.10 volt and in position is taken as ± 2.0 cm.

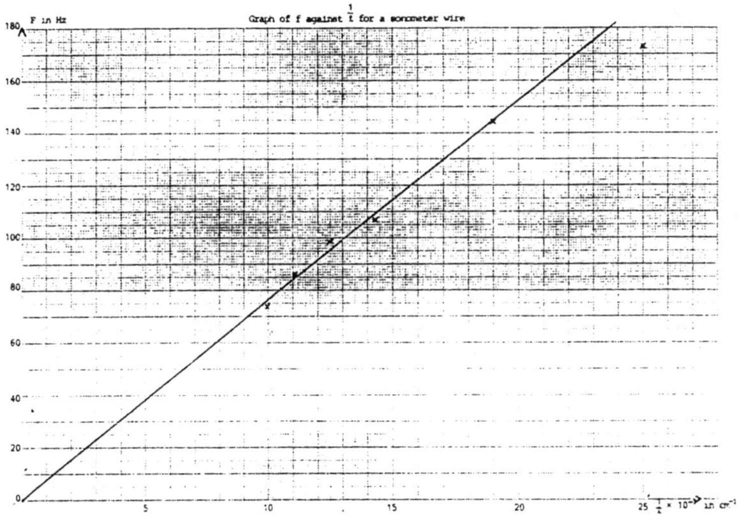

FIG. 3. Graph of f against $1/l$ for a sonometer wire.

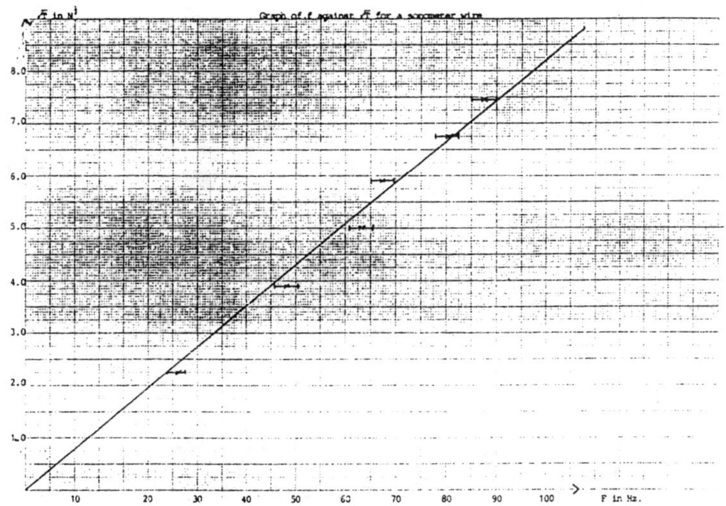

FIG. 4. Graph of f against $\sqrt{T}$ for a sonometer wire.

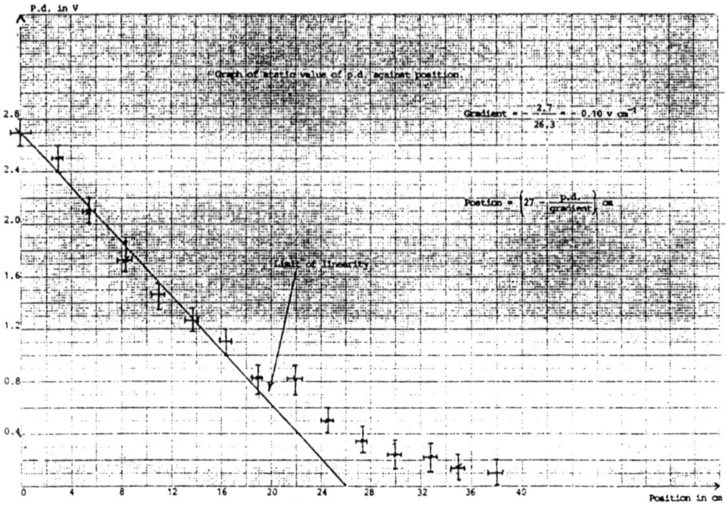

FIG. 5. Graph of static value of p.d. against position.

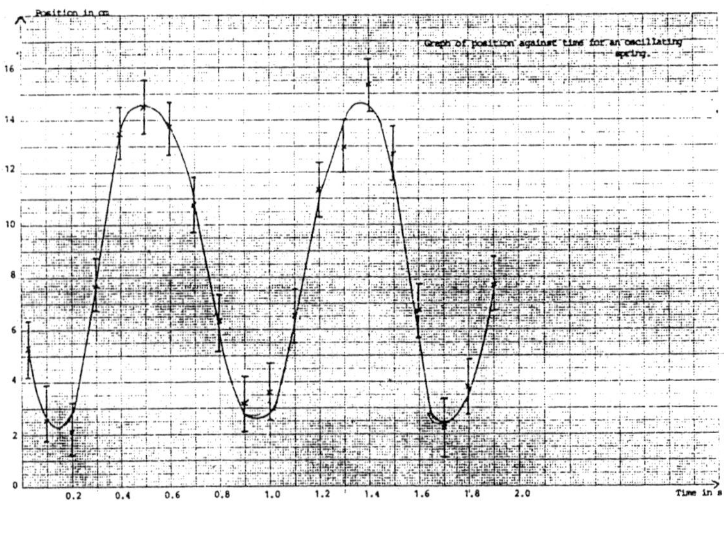

FIG. 6. Graph of position against time for an oscillating spring.